MEN AND MINES OF NEWMONT

Men and Mines of Newmont

A FIFTY-YEAR HISTORY

by Robert H. Ramsey

OCTAGON BOOKS

A DIVISION OF FARRAR, STRAUS AND GIROUX

NEW YORK 1973

Copyright © 1973 by Newmont Mining Corporation
All rights reserved
First printing, 1973
Library of Congress catalog card number: 72–6985
ISBN 0-374-96710-5
Printed in the United States of America
Published simultaneously in Canada by
Doubleday Canada Ltd., Toronto
Designed by Dorris Huth
Drawings by Clifford Schule

The drawing for chapter 1 is adapted from a photograph in *Bonanza West* by William J. Greever, published by the University of Oklahoma Press. Additional acknowledgment is made to the Montana
Historical Society.

The drawings for chapters 2 and 4 are based on pho-
tographs published in *Gold Cities* by Robert M.
Morley and Doris Foley, courtesy of Mr. Morley. For
the drawing for chapter 4, acknowledgment is also
made to the Nevada County Historical Society

Contents

MEN AND MINES OF NEWMONT

W. B. Thompson's office, 1921

PROLOGUE: The Leaders of Newmont

On Monday, May 2, 1921, New York enjoyed a cool, though overcast, day, free of the storms that had troubled the city a few days previously. In his corner office at 14 Wall, fifteen floors above the Street, Colonel William Boyce Thompson clipped and lighted his third cigar since lunch. Smoking too much, his doctors said, but a man had to hang on to some of the simpler pleasures.

Preceded by a discreet knock, Judge Ayer came into the office, his face set in its usual mask of calm efficiency. Carefully threading his way between the jaws of the black bearskin spread on the floor at his right and the grizzly bear's snarl at his left, Ayer reached the big desk and said, "I've heard from Wilmington, W.B. They held the first meeting of Newmont Corporation this afternoon, and they took care of all the formalities. I've made the necessary announcements."

"Good," said Thompson. "Have you set up the board meeting for tomorrow?"

"Yes, sir," said Ayer. "At eleven o'clock here in the office. We'll elect Ted Schulze, Thomas, and Dodge as directors, then elect Schulze, president; Thomas, vice president; and Dodge, secretary and treasurer. Just as we agreed. Mackenzie will be Dodge's assistant."

"All right, then," said Thompson, his moon face impassive, "just make sure Ted is on time."

Permitting himself a slight smile, Ayer nodded, turned, and retraced the narrow path between the threatening bears. As the door closed behind him, Thompson settled his bulky frame more comfortably in the big chair and blew a cloud of smoke up at the cheerful countenance of the head of the bull moose on the wall to his left. Bear or bull? Today the colonel felt remarkably bullish.

Flamboyant, imaginative promoter that he was, Thompson had running through his mind that afternoon one grandiose plan after another involving his new company, Newmont Corporation. Yet one doubts that in any of his plans he looked ahead for fifty years, or even twenty. In Thompson's world, growth was measured in days, not years. A trader at heart, Thompson wanted action, and through this new company he hoped to get it on a massive scale and through one quick turnover after another.

Yet fifty years later, the company he founded owes much to Thompson's foresight in building into its structure a mechanism for solid growth as a mining company. Had he maintained Newmont as only a trading vehicle for his many promotions, however far-reaching they may have been, it is unlikely that Newmont would have long survived him.

As it is, Newmont Mining Corporation (Thompson added the word "Mining" in 1925), represents an achievement vastly greater in terms of geography, diversity, productivity, earning power, and asset value than even Thompson's wildest dreams could have pictured.

He organized Newmont in 1921 with a capitalization of $8 million. At its maximum in 1971, its fiftieth year, Newmont's approximately 24 million common shares had a market value of $942,350,000, and the estimated fair value of its gross assets was $950 million. Certainly, on that afternoon in 1921, Thompson could hardly have thought in terms so huge, or of times so remote as 1971, yet his thoughts did lead with amazing clarity to the very road that Newmont has so successfully followed.

Thompson's basic formula was simple enough: Find potentially profitable mineral properties; form companies to finance, develop and manage them; *promote the sale, usually to the public, of new issues of stock in these companies;* and build a liquid, income-producing portfolio of mining and petroleum securities to use in financing still greater mining enterprises.

Delete the clause in italics, and you have a simple statement of the philosophy of Newmont today. However, the major differences between Thompson then and Newmont now lie in the application of this philosophy. Thompson was a trader, a fast mover. Newmont neither promotes companies, nor does it trade actively. Thompson never really managed a mining property. He was too impatient, too hungry for results. Newmont does manage mineral properties, eleven of them at present, with major investments in a dozen or so more. Yet the thread of Thompson's basic idea runs all through Newmont's history, and it is clearly visible today.

To have lasted so long and succeeded so well, the idea must have had merit, but it could not have survived for long without the second basic ingredient of the Thompson formula—the right men. Early in his career, Thompson learned that even more important than money were skillful, intelligent men, the best men he could find. From Thompson on, Newmont has always sought out the men most qualified for the job in hand, and the company has enjoyed a high degree of success in finding them, as its history shows.

Not that Newmont is unique among mining companies in this respect. Most mining company administrators are good at their jobs because they have to be. However, for many years, Newmont was unique in that a very small group had to make, continuously and rapidly, a multitude of critical decisions covering every possible phase of the mineral industry. Although today Newmont's managerial group is somewhat larger, its responsibilities have grown, too, now covering twenty or more operations spread all over the world and clamoring for attention all at once. Obviously, a few successive wrong decisions could quickly slow down, or wreck, the whole enterprise, and it becomes vital, more so than in most mining companies, that critical decisions be made rapidly, firmly, and correctly.

This is not to say, however, that Newmont's men never make mistakes. In this chronicle of the company's first fifty years, there will appear no supermen, no infallible administrators. From the very first, the men directing Newmont have suffered the same fears, doubts, and worries common to all of us. They have disagreed among themselves, sometimes violently, and they have made mistakes, some of them large and expensive mistakes, which appear in these pages along with their successes.

But in sum, Newmont's men have been right much more often than they have been wrong. Their doubts and fears have served not to weaken their efforts but to strengthen them by calling forth the intensive study required to dispel such doubts. Furthermore, there has always existed in Newmont a rare kind of camaraderie that rises above even deep disagreements and creates a warmly effective atmosphere in which to work.

How remarkable that so simple a corporate philosophy, and so exacting a management technique could have been maintained so successfully over fifty years by four men so utterly unlike each other as the four chief executives of Newmont Mining Corporation: William Boyce Thompson, Charles F. Ayer, Fred Searls, Jr., Plato Malozemoff—four men widely dissimilar in personality, appearance, dress, tastes, interests, and knowledge, yet remarkably alike in the cool and accurate judgment required to guide a multifaceted enterprise like Newmont.

Actually, as the imaginary dialogue that opens this history

indicates, Newmont's first president was Theodore Schulze, Thompson's son-in-law. When he formed Newmont Corporation in 1921, Thompson wanted a vehicle for his various promotions of greater power and scope than the old Newmont Company he had incorporated in Maine in 1916. He also, as an adoring and indulgent father might, appears to have tried to establish his daughter Margaret's husband, Ted, in a business career. Whatever the reason may have been, the assignment was short-lived. Unqualified by experience or talent for Thompson's high-speed operations, Ted Schulze resigned as president in July, 1923. Charles Ayer replaced him as president and director.

During the twenty-three years of Ayer's tenure as president, Thompson's control or influence, all-powerful at first, slowly declined until his death in 1930, and in 1931 management responsibilities were given to a triumvirate of Ayer, Franz Schneider, and Fred Searls, which ran the company until January, 1947, when Ayer became chairman and Searls, president. In January, 1954, Malozemoff became president and Searls, chairman.

Reserving Franz Schneider for a later chapter, the other four men make a striking comparison. Thompson was of medium height, corpulent, bald, his face moon-shaped and usually expressionless. Ayer was short, clean-cut, impeccably dressed, the very image of what used to be called "a gentleman of the old school." Searls, also short, but roughhewn in countenance, had an athlete's build; in fact, he earned quite a reputation as an amateur boxer. An eccentricity in dress, befitting his personality, was his trademark. Malozemoff, slight in stature, thin-faced, smooth-haired, appears cast in the mold of today's alert, intelligent, conservatively dressed executive.

Do the times make the man? Newmont's experience might argue for the man growing to meet the times. Thompson's promotional skill, Ayer's cool, accurate judgment, Searls's hard-driving leadership and host of friends, Malozemoff's sophisticated financial skill and incisive mind—all seem designed remarkably well for the specific period in which their influence has been strongest.

Thompson, the organizer and promoter, tended to flamboyance in his approach to business. His office paraded his hunting
trophies, from the bearskin rugs before his desk to the genial
moose up at his left, and the enormous brooding buffalo head
on the wall over his right shoulder. Incidentally, Thompson's
massive, beautifully carved desk is still in the family, currently
being used by Dave Pearce, Newmont's vice president for operations. The animals disappeared long since.

In his handling of his own and Newmont's affairs, Thompson reserved all critical decisions for himself. However, he depended heavily for guidance on his staff, and no mining industry
leader of his day had a better one. His first group, prior to
Newmont, included George Gunn for prospecting, Henry
Krumb for meticulous mineral property evaluation, Walter Aldridge and Wilber Judson for engineering, H. F. J. Knobloch
and Charles Ayer for legal matters, A. J. McNab for copper
smelting, Dave Thomas, Henry Dodge, and Frank Holmes for
finance and accounting.

Aside from these men and others he later brought to Newmont, Thompson's greatest contribution to the company lay in
the mineral properties to which he gave Newmont access:
Magma Copper Company, for example, and Kennecott, and
Texas Gulf Sulphur. Thompson supplied the company name,
"New" from New York and "Mont" from Montana, because he
grew up in the latter and made his money in the former. It was
his money that laid a solid foundation for Newmont, which
years later stood the company in good stead. Equally important
was his host of friends, from Thomas W. Lamont of J. P. Morgan & Co., to Doby Tom, his man of all work in Arizona. All
these elements helped Thompson and helped Newmont, but
until the late 1920's, when failing health forced him to curtail his
activities, Thompson remained the strongest influence in forming the decisions by which Newmont stood or fell.

Charles F. Ayer's highly trained lawyer's mind rejected any
attempt he, himself, might have considered to try to duplicate
Thompson's authoritarian control. However, it would be a mis-

take to underrate Ayer's quiet yet firm guidance, and throughout his tenure, Ayer remained as Newmont's court of last resort. Because of this, and because of his unquestioned legal competence, he became known as Judge Ayer, a nickname that bespoke his associates' great respect.

Nevertheless, beginning with the advent of Fred Searls in 1925, one can detect a spreading of the decision-making process to other members of the Newmont hierarchy. Through the 1920's and 1930's, Newmont acquired a management group of talented, strong-minded, individuals who would not have taken kindly to authoritarian management. Fred Searls, himself, was as independent a thinker in geology as ever lived. Remaining from the Thompson era was A. J. McNab, a tough-minded, sometimes stubborn, Scotsman. There was Phil Kraft, urbane, pleasant, but not to be led willy-nilly by anyone. Franz Schneider stood firmly as the authority on finance and investment. Ayer, Fred's brother, Carroll Searls, and young Roy Bonebrake took care of legal matters. Lastly, Henry DeWitt Smith, a newcomer, had grown into a tower of strength for Newmont in mining engineering.

All these men were strong, intelligent, and capable. They formed opinions on everything, and they could defend them. Seldom were they all in agreement, unless they felt Newmont to be under attack from the outside. Consequently, there developed a kind of horizontal organization, wherein responsibility split up along geographic lines. Fred Searls handled exploration and new projects generally. Smith's primary responsibility was in mining operations, first in Canada and then in Africa with O'okiep and later, Tsumeb. McNab took care of Magma as operating head under Ayer as president and made clear that interference was unwelcome. Phil Kraft supervised Canadian operations and Newmont Oil Company. Judge Ayer, Henry Dodge, Franz Schneider, Carroll Searls, and Roy Bonebrake strove for harmony, solvency, and legality, with marked success, except for an occasional lapse in harmony. It was essentially this same group of strong individuals who directed the company through the 1940's.

With Malozemoff's accession to the presidency in 1954, a change in management structure that had been quietly developing became more evident. The old horizontal division, or geographic split, began to disappear. M. D. Banghart, who had been manager of various operations in Canada and Africa, was brought to New York and succeeded Henry Smith in charge of all Newmont's production operations. Frank W. McQuiston, Jr., with Newmont since 1935, supervised all Newmont's metallurgical developments. E. H. Tucker, as chief engineer, was, and is, on call by any Newmont operation. Bob Fulton, brought in by Fred Searls, became responsible for much of Newmont's exploration activities. Roy Bonebrake became executive vice president and chief counsel.

These men, and their successors, have made up a group that may shoot fewer sparks than the old group did, but it has set a performance record even more impressive than the great accomplishments of Newmont's past.

In recent years, there have been further changes, in general looking toward a stronger organization along functional lines, indicating Plato Malozemoff's awareness of the need for surrounding himself with men competent in the special skills a mining company lives by. True, Malozemoff is utterly unlike his predecessors in philosophy, attitudes, and ideas of organization, but he has not deviated in his drive for growth as a *mining company* within the broad area of the mineral industries.

Furthermore, as his development of competent, specialized staff indicates, Malozemoff tends to depend much more heavily than his predecessors on a critical study and analysis of a given situation, rather than on intuition or a highly personalized approach. Newmont's administrators today are all professionals with highly developed skills in financing, exploration, mining, metallurgy, and engineering, plus a joint demonstrated ability to anticipate major economic and political changes.

How is it possible that with so varied a leadership, so complex a collection of affiliated enterprises, and so strong and, at times, so temperamental a set of management groups, Newmont could yet show remarkably steady growth over a turbulent fifty-year period without losing its essential direction or character?

The answer appears in the following pages, wherein it is hoped is recorded as accurately as anyone outside the events can ever record them, the lively history of Newmont's first half-century.

Gold miner, Alder Gulch, 1869

1 The Flamboyant Promoter

Although Newmont Mining Corporation was born in the brain of a man occupying a Wall Street office early in 1921, for a real understanding of the company's origin, one must go back over a century to the tiny, but uproarious gold camp of Alder Gulch, Montana, where William Boyce Thompson was born on May 13, 1869. An offshoot of the main camp at Virginia City, where gold had first been discovered in Montana, Alder Gulch was a fascinating place for a healthy boy to grow up. Thompson did not just read about prospectors, Indians, and road agents—he grew up with them. Born and bred in the rough-and-ready life of the frontier, Thompson kept, throughout his life, a fondness for the out-of-doors, the western United States in particular.

Shortly before Thompson reached high school age, his father gave up on Alder Gulch and moved his struggling lumber and carpentry business up to Butte, Montana. That ugly, sprawling mining camp gave the young Thompson certain solid assets that stood him in good stead throughout his life. Foremost was an acquaintance with the mining industry, particularly with the copper phase of it. Second was a thorough knowledge of human nature, at least those aspects of it exhibited in the exuberant and sometimes violent atmosphere of Butte. Third was an ability to play poker, at which he became outstandingly skillful. Finally, and possibly most important, was his contact with an English teacher named Arthur Newell, an Oxford graduate who in some mysterious way found himself in charge of Butte's only high school. Newell recognized in the young Thompson an unusually keen intelligence, and it was owing to Newell's insistence that Thompson's parents made the sacrifice of sending him east to Phillips Exeter Academy in New Hampshire.

A man who has experienced an Eastern preparatory school education seems inevitably, at least in some degree, to be marked by it for life. Thus, Exeter had a strong effect on Thompson, not just because of what he learned there academically but because of the friends he made and the vision it gave him of horizons wider and more fascinating in their attraction than were ever conceived in Butte.

Possibly the most influential of Thompson's Exeter friendships was the one he formed with Thomas W. Lamont, who was later to become a partner in J. P. Morgan & Co. As extraordinary in his way as Thompson, Lamont was a country boy from a farm in upstate New York. Although quite different in tastes and background, Thompson and Lamont shared a driving ambition, and each young man saw in the other certain qualities that he admired. The friendship that began at Exeter lasted throughout their lives.

Thompson did not graduate from Exeter, but he stayed there long enough to acquire enough credits to enter the Columbia College of Mines in New York. Thompson's stay at Columbia was both short and undistinguished. He was much more interested in all-night poker sessions than in the classroom. Also, a couple of trips to the visitors gallery at the New York Stock Exchange gave him an idea, which came to the surface only years later, that the real interest and the real reward in mining lay on Wall Street and not in the mines of the West.

Giving up on Columbia after a year, Thompson went back to Montana and began a series of attempts to open up his own mining property. One such venture was successful only in that it brought Thompson the acquaintance of a self-taught engineer and geologist named George Gunn. As it had happened with Lamont, Thompson and Gunn discovered themselves kindred spirits in their fondness for poker, for mining, and for the outdoors generally. Gunn worked closely with Thompson thereafter until Gunn's death in 1913.

In February, 1895, Thompson married a childhood friend, Gertrude Hickman, and the couple settled down in Helena, Montana, where Thompson engaged in a lumber and coal business, which was only moderately successful. Thompson also essayed a venture into Montana politics, but neither the voters nor the party leaders were impressed by his appearance or his personality. The poker face that was his asset at cards was a liability with the voters, and his thinning hair and expanding waistline had already given him an air of smug solidity that alienated the politicos.

From lumbering and politics, Thompson turned to managing an investment company, formed with his brothers and his father. This activity held his interest for a short while, but after two years of increasing boredom, Thompson began to think more and more wistfully of the opportunity he had seen in the surge of activity surrounding the New York Stock Exchange. Suddenly the hidden, flamboyant promoter broke through the skin of the unemotional small businessman. To the dismay of his family, Thompson, in January, 1899, packed a satchel full of mining stock certificates, gathered his wife and their daughter, went off to New York, and checked into the Waldorf-Astoria, which he remembered as the best hotel in town. Here he was, thirty years old, a family man, supported only by a series of unsuccessful mining promotions, possessor of nothing of any intrinsic value except a pitifully small cash reserve, yet occupying an expensive suite of rooms with all the aplomb of the magnate he was one day destined to be.

For about a year, things looked pretty grim for the Thompsons. Some sales of stock in Boston kept them going, but not until an opportunity arose to promote an actually operating mine did their fortune change.

Destiny in the shape of the Shannon Mine near Clifton, Arizona, now took a hand. Hearing of the Shannon Mine on a visit to Boston, Thompson went out to see it for himself. He liked what he saw of the mine and, furthermore, fell in love with the Arizona country itself, an affection that he never lost. Thompson then acquired the Shannon Mine through a bit of expert financing that was his first taste of what courage and skill in mine promotion could do. To run the property, Thompson turned to an acquaintance of his from Butte, who was reputed to be a good mine manager. This man, Philip Wiseman, agreed to take on the Shannon, thereby giving Thompson his first successful mining venture and establishing an acquaintance that proved even more valuable years later. Although Thompson's connection with the Shannon was rather short-lived, he came out of it with a profit of some $75,000, and in 1904, he approached his old friend, George Gunn, with the invitation to look for a new mine.

At about this time the whole nature of the copper mining industry in the United States was radically altered by Daniel C. Jackling's conviction that a copper deposit, of what seemed then to be of an uncommercially low grade, could be mined profitably by means of large-scale open-pit operations using steam shovels and rail haulage. The result was the famous Utah Copper development at Bingham Canyon near Salt Lake City. In the spring of 1906, a small steam shovel took the first bite out of what has since become the largest man-made excavation in the world.

Thompson, through Hayden, Stone & Co., helped raise $1.5 million as part of the financing of the original 6,000-ton mill and thus became thoroughly familiar with the project. He also became acquainted with a young geologist named Henry Krumb. When Jackling invited the Guggenheim family, founders of American Smelting and Refining Company, to inspect the Bingham Canyon orebody, the Guggenheims, sensing a profitable investment, asked their chief engineer, John Hays Hammond, to examine the property. Hammond turned the actual job over to Henry Krumb. Krumb had performed well for the Guggenheims on other properties, and Hammond had great faith in his judgment.

Later, Henry Krumb was identified with Newmont Mining through nearly all his working life. Born in Brooklyn of parents of Pennsylvania-Dutch ancestry, Krumb was destined by his family for his uncle's bar in Brooklyn but sought escape and found it in the form of a scholarship in the Mining Department at Columbia College in New York. His graduation as a mining geologist benefited both Krumb and the college, for in his will Krumb directed that most of his estate, about $18 million, was to go eventually to the Columbia College of Mines.

Among Henry Krumb's outstanding characteristics was a passion for correctness in detail of whatever activity in which he felt it worthwhile to engage. For example, in his work of drilling and sampling the Bingham Canyon orebody, he employed sixteen junior engineers and kept them busy for seven months. This job laid the foundation of Krumb's expertise in evaluating porphyry coppers, and the methods he developed later at Ray, Arizona, were eventually adopted wherever geologists faced similar sampling jobs. In fact, so reliable was the Krumb technique that no one to this day has discovered a way to improve on the methods he developed. For example, in 1946, Magma Copper Company used Henry Krumb's drilling and sampling technique to delineate the San Manuel orebody in Arizona. It is not too much to say that Henry Krumb was the first man to apply scientific methods to evaluation of these porphyry copper orebodies.

Because of his acquaintance with Jackling and the Utah Copper Company, Thompson could observe Henry Krumb's skill and initiative at close range. Not one to stand on ceremony,

Thompson at once offered Henry Krumb a job on his small staff. Krumb refused this offer, but Thompson did not give up easily, and years later he won Krumb over as a member of his personal staff.

In the meantime, Thompson and George Gunn had made a killing on a copper prospect near Ely, Nevada. A group headed by a young engineer named Mark Requa had uncovered a copper property there, which George Gunn thought had possibilities. At Gunn's suggestion, Thompson staked claims in the surrounding area covering certain copper showings, but more importantly, including water rights that any mining or smelting operation in the area would have to draw on eventually.

By 1906, the Requa group had formed a company called Nevada Consolidated and was selling stock. Thompson was able to get an option on 40 per cent of the stock for $12.00 a share, which he thereupon offered to the Guggenheims at $12.50 a share. Again, the Guggenheims called on Henry Krumb, who examined the Nevada Consolidated property and recommended the purchase. The Guggenheims bought Thompson's shares, thus yielding him an immediate profit of $200,000. This in an age that had yet to hear of income or capital gains taxes.

Thompson and Gunn then organized the Cumberland Ely Copper Company and sold 51 per cent of it to the Guggenheims on the strength of Cumberland Ely's vital water rights. This resulted in the ownership by the Guggenheims of 40 per cent of the Requa, and 51 per cent of the Thompson-Gunn Company. Following considerable, and rather heated, negotiations, the companies were merged under the name of Nevada Consolidated, minus both Messrs. Requa and Thompson. Nevertheless, Thompson and Gunn came out with cash and stock worth about a million dollars.

To the thirty-seven-year-old Thompson, the evidence was clear that the way to get rich was to promote mining stocks and trade in them. A lesser man might have yielded to a temptation to promote mining properties that were something less than worthwhile, but there is nothing in the record to indicate that

Thompson even felt such temptation. He had too great an affection for mining and a real respect for the productivity of the earth to become involved in anything that looked like a swindle. Besides, when there were so many rich orebodies awaiting discovery, a man would be a fool to risk promoting a worthless mineral property.

The years after 1906 were remarkably successful for Thompson. His reputation as a promoter grew rapidly, and gone long since were the days of pretending to an affluence he did not have. Thompson no longer had to look for opportunities; people brought opportunity to him.

Typical of such an opportunity was the Nipissing Mines Company promotion, one that aroused world-wide interest and nearly precipitated a disastrous financial storm. Discovered by accident late in 1903 at a point along the railroad about halfway between Toronto and Hudson Bay, the Nipissing ore was reputed to be unbelievably rich in silver and cobalt. A New York metal broker, E P. Earle, and a group of friends acquired claims in the area for $200,000 and then organized the Nipissing Mines Company, capitalized at $6 million with 1.2 million shares outstanding of a par value of $5.00 each, all on the strength of a few samples of almost incredible richness.

Early in 1906, there developed what years later turned out to be a fairly frequent pattern. The owners, the Earle group in this case, approached Bankers Trust in New York in a search for funds. There they were referred to Thomas W. Lamont, who recommended W. B. Thompson as a man who could help them. Lamont, who had not a shred of flamboyance in his make-up, had quickly recognized the value of Thompson's promotional skill combined with his basic integrity. The two had become even closer friends than they had been at Exeter.

Recognizing at once the possibilities in Nipissing, Thompson first took an option for himself on 240,000 shares at $3.40 with the right to buy 360,000 shares more at $4.90. Thompson then set about promoting the property. The job could hardly have been easier. A minimum of work on the claims disclosed a vein 6 to 28 inches wide, 600 feet long, apparently mostly cobalt and silver. Guards were posted round the clock on this "Silver Sidewalk." The thought of wealth wide enough to walk on caught the public fancy. First offered to the public in May, 1906, at $4.00 per share, the price of Nipissing rose to nearly $12.00 by September. Then news of an even richer strike at Nipissing broke over the market, and by the end of September, "Nip" hit $20.50. By then, Thompson had sold 300,000 of the 600,000 shares he had purchased on his original option. George Gunn hung on to a lesser number of shares he bought in the early promotion.

Thompson, himself, really believed in the property he was promoting, and to insure a greater stability for Nipissing he suggested to the Guggenheims that they buy the 400,000 shares of Nipissing still held by Earle and his group. The Guggenheims sent John Hays Hammond, the most respected mining engineer of his time, up to have a look at Nipissing, and Thompson and Gunn went up with him. While Hammond was examining the outcrop, Gunn made his own study of it, after which he came back to Thompson and quietly asked him to sell his Nipissing shares when Thompson got back to New York.

Obviously, Gunn had not liked what he had seen, but Thompson felt it was no time to ask further questions. Concluding that he had better keep Gunn away from Hammond until the latter had reported back to the Guggenheims, Thompson challenged Gunn to a game of Canfield. Gunn accepted, and the two concentrated on their card game throughout their train trip back from Nipissing. Gunn left the party to head west without ever talking again with Hammond. Instead of waiting until he reached New York, Thompson sold Gunn's shares by telephone from Ottawa, and later sold a large block of his own shares at a nice profit for both men.

Acting on Hammond's favorable report, the Guggenheims took an option on 400,000 shares at $25.00 a share, paying $2.5 million down as the first of four installments. News of the Guggenheims' action pushed the stock up over $33.00. How high

Nipissing might have gone is anyone's guess, but before long, news began to leak out that the famous "Silver Sidewalk" was pretty much on the surface and narrowed drastically in depth. Gunn, with his practical training, had seen the truth that Hammond had missed. Nipissing began to slide, and Thompson sold the last of his holdings. Remember, this was 1906, when this sort of activity was considered not only proper but quite admirable.

The Guggenheims decided to drop their option, and when this news became public, Nipissing plunged. Widespread criticism of the Guggenheims caused them to offer to refund any losses borne by investors who had bought Nipissing through their company, but the stock still declined. In the end, although Nipissing turned out to be less productive than the original promotions had anticipated, neither was it ever a total failure. Thompson later bought back into Nipissing, and the Cobalt area in Canada is still in production. Nipissing, itself, continued for years to yield a reasonable income to Thompson and those others who joined his re-entry into the company.

The main result of the Nipissing venture, however, was that Thompson wound up with about $5 million in cash plus the experience of a promotion that made him more than ever eager to take on new assignments. This new-found confidence was quickly put to a severe test by the panic of 1907, which came close to wiping out all of Thompson's newly acquired fortune. The reversal only stiffened Thompson's resolution.

Late in 1907, Henry Krumb brought Thompson an opportunity to recoup. Krumb had been sent by the Utah Copper group to Ray, Arizona, where he drilled and sampled another porphyry copper orebody. While there, Krumb was invited by Phil Wiseman to come up to Globe and look over some claims known as the Inspiration prospect, held by a group of men in Kansas City. Krumb was quite interested and approached Daniel Jackling, who also thought well of it. However, Charles Hayden, the principal financier of the group, objected to the terms, and a deal with the Utah Copper Company fell through.

Krumb then got a one-week option on the Inspiration property on his own and took it to George Gunn and Thompson.
Years later, Thompson might have acted solely on Krumb's opinion, but at that time, before deciding, he sent Gunn to look at
the Inspiration. Because a decision had to be made quickly,
Thompson set up a telegraphic code for Gunn's use. Shortly
after Gunn reached the Inspiration claims, he sent back a telegram to Thompson containing the one word "Betty." Of Gunn's
several girl friends, Betty was the one for whom he had the
greatest affection, and immediately Thompson sent Krumb and
Charles F. Ayer, his new lawyer, to Kansas City to negotiate a
deal with the owners of Inspiration.

Thompson got the Inspiration, as he got most material
things he wanted, and with it, at long last, he got Henry Krumb.
For the rest of his life, Krumb remained loyal to Thompson and
later to Newmont, although he steadfastly refused to go on the
Newmont payroll, either as an officer or a staff member. He did
consent to become a Newmont director, and as such, served the
company well, but he never worked directly for Newmont except as a consultant.

Also in 1907, Thompson made another major acquisition in
the person of Charles F. Ayer, mentioned above, a young California lawyer anxious to try his luck in New York. By chance,
Ayer had moved into a house in Brooklyn Heights near the one
occupied by H. F. J. Knobloch, Thompson's lawyer, and the two
became friends.

When Knobloch wanted to take a vacation in the summer
of 1907, he suggested Ayer to Thompson as his replacement.
There still exists in the Newmont files the yellowing fragile document, signed by W. B. Thompson and Ayer, by which Ayer
promised to work full time for Thompson on any and all legal
matters at a salary of $3,500 per annum. Ayer worked on the In-
spiration negotiation as his first big assignment, and after Knobloch went to Texas Gulf Sulphur in 1918, Ayer handled all
Thompson's legal affairs.

So Thompson formed Inspiration Copper Company in

1908, with Krumb as consulting engineer and director of the drilling and sampling program. When it became clear that sufficient ore was present to justify production, Thompson put Walter H. Aldridge in charge of planning the mine and the surface plant. Aldridge for fourteen years had been in charge of the mining and smelting activities of what is now Consolidated Mining and Smelting Company of Canada at Trail, B.C. Thompson had persuaded Aldridge to join him in Arizona, which he did on January 1, 1911. Aldridge was followed not long after by Henry E. Dodge, A. J. McNab, and Wilber Judson, all from Cominco, and all brilliant men.

Shortly after Inspiration was incorporated, the company was consolidated with the neighboring Live Oak copper property owned by the Anaconda Company. Late in 1911, a new company was formed called Inspiration Consolidated Copper Company, which is still in production and in which Anaconda retains a 28 per cent interest. Thompson was the first president of Inspiration, and William E. Corey, president of United States Steel, and Albert H. Wiggin, president of the Chase Bank, were both on the first Inspiration board of directors. Both men were close friends of Thompson's, and Wiggin became a member of the Newmont board in 1925 where he served actively until 1949. Newmont retained its stock interest in Inspiration until 1924.

Not all Colonel Thompson's promotions turned out to be Inspirations and Nipissings, however. Proof that the colonel, too, could make mistakes lies in the history of Mason Valley Mines Company, which he organized on January 4, 1907, to work a copper property near Yerington, Nevada.

A spectacular outcrop of Mason Valley's contact metamorphic orebody had misled a number of predecessor companies into trying to mine it. Some ore shoots actually ran 65 per cent copper, and there were outcrops of almost pure malachite that had been mined and sold as bluestone. Thompson was quite taken with it, but Henry Krumb was not bemused by rich samples and gorgeous outcrops. He told the colonel there wasn't enough ore to make a worthwhile mine and recommended against it. Thompson, for once, overrode Krumb's judgment, and put Mason Valley into production, with George Gunn as manager, based on the idea of mining and smelting Mason Valley's ore together with ores from other small mines as far away as California.

Mine development, through drifts and raises, confirmed Krumb's judgment; reserves were not impressive and the average grade was about 3.5 per cent copper. Nevertheless, Thompson had McNab design and build a smelter, including two blast furnaces and two converters, that cost about $900,000 and had a capacity of 1,800 tons of ore per day. The smelter was blown in during January, 1912, and produced about 8,500 tons of copper that year.

From such records as are still available, it appears that operations were quite profitable in 1912 and 1913, but Mason Valley showed a loss of $40,275 for 1914, reducing its net surplus to $307,500. Feed to the smelter by that time was half Mason Valley ore and half custom ore, on which high freight rates were ruinous. The smelter shut down in October, 1914, and resumed in February, 1917. Profit in 1917 was $358,277, but it dropped to $178,202 in 1918 and disappeared entirely in 1919, when the price of copper collapsed. From March, 1919, until early 1926, Mason Valley was inoperative.

In the meantime, Thompson equipped Mason Valley with a flotation mill to try to get more copper through the smelter, and he tried hard to arrange lower freight rates for the custom ore on which Mason Valley depended. Neither effort was successful, and Thompson had at last to give up. The company was dissolved as of April 22, 1929.

Aside from its prominence as one of Thompson's less successful efforts, Mason Valley is important historically for two reasons: the men who ran it and the investment policy it initiated.

At one time or another, all of Thompson's staff was involved with Mason Valley. Aldridge, Judson, Knobloch, Dodge, and McNab, all were officers of the company, but only Dodge and McNab stayed with Newmont. Also, when McNab was running Mason Valley during its best period of 1917–18, he had George A. Kervin as smelter superintendent, an association that greatly benefited Newmont in 1929 and again in 1937 when Kervin directed the exploration, smelter design, and initial production at O'okiep in South Africa.

Mason Valley's investment policy was, as far as can be determined from available records, the first instance of Thompson's use of a mining company as a sort of holding company for other investments. Mason Valley used most of its funds to buy stock in other mining companies, most prominent of which was its purchase in September, 1916, of nearly 100 per cent of the stock of Gray Eagle Copper Company for $271,250. Gray Eagle owned, but wasn't working, a small high-grade copper orebody near Happy Camp, California. Newmont eventually became owner of Gray Eagle, and mined out the orebody during World War II.

Other details on its investments aren't available, but Mason Valley did well enough on its portfolio to be able to report profits from 1921 to 1926 even when both mine and smelter were shut down. These reached a high of about $84,000 for 1925. It could be that this experience guided Thompson in setting a similar, but much broader, policy for Newmont Mining Corporation. If so, for the men it brought in and the policy it helped establish, Mason Valley was well worthwhile.

Success with Inspiration led to Thompson's involvement in dozens of enterprises. One of the larger and more successful of these was Texas Gulf Sulphur Company, which Thompson helped organize in 1918. Ten years or so later, Newmont Mining Corporation turned a neat profit on some of the TGS stock Thompson thus acquired.

Interest in sulfur in Texas had been aroused back in the 1880's, but nothing really significant had developed until the Frasch Process began to prove successful shortly before the First World War. The inventor of the process, Herman Frasch, was born in Württemberg, Germany, on Christmas Day in 1851. He came to the United States at the age of nineteen and found a job in a Philadelphia chemistry laboratory through which he quickly developed an interest in the petroleum industry in Pennsylvania.[1]

Frasch was one of those geniuses whose brain produces a stream of original ideas. Becoming interested in the problem of removing sulfur from Ohio crude oil, he developed a lucrative process for doing so, and went on to develop a keen interest in the element itself. Discovering that ordinary underground mining methods used in Texas could never compete with the low-cost, high-grade Italian sulfur mines, Frasch cast about for a cheaper way to get the sulfur up to the surface. It occurred to him that inasmuch as sulfur melted at 114 degrees centigrade, it might be possible to pump superheated steam underground to melt the sulfur and allow the molten material to be pumped back up to the surface.

Beginning in about 1893, Frasch began to work on developing his process. The string of discouragements he encountered would have frustrated many inventors, but like Thompson, Frasch had an unshakable confidence that carried him through every disappointment.

To shorten the story, Frasch was, of course, eventually successful, and one of his greatest successes took place on Big Hill in Matagorda County, Texas. In 1908, a group of St. Louis men organized the Gulf Sulphur Company to work claims on Big Hill, but until this company attracted the attention of Seeley Mudd and Bernard Baruch, it never made much headway. Mudd and Baruch saw possibilities in Big Hill, and in 1916 they began negotiating with the St. Louis group to acquire their interest. To assist them, Mudd and Baruch approached J. P. Morgan, who, in turn, brought in Thompson.

Obtaining control of Gulf Sulphur Company in 1916, Thompson's group actively began development of sulfur produc-

[1] Information from *The Stone That Burns* by William Haynes, copyright 1942 by Litton Educational Publishing, Inc. Reprinted by permission of Van Nostrand Reinhold Company.

tion at Big Hill. Because of the confused ownership of certain mineral rights on the property, Thompson sent his lawyer, H. F. J. Knobloch, down to direct the project. Knobloch unraveled the legal tangles quite successfully and production began. In 1918, the name of the company was changed to Texas Gulf Sulphur Company, by which name it was known until 1972, when the name became Texas Gulf Inc. Knobloch sensed great possibilities in the company, and decided to stick with it, becoming treasurer, a position he held until his retirement in December, 1947.

When the United States entered the war in 1917, Seeley Mudd resigned as president of Texas Gulf to take part in the war effort, and Walter Aldridge, Thompson's chief engineer, was elected president in his place. As happened with Knobloch, Texas Gulf Sulphur caught Aldridge's interest, and he remained as chief executive officer until his retirement in June, 1951. In addition to these two, Thompson found himself supplying the operating vice president for Texas Gulf Sulphur in the person of Wilber Judson, who for several years had been a mining engineer on Thompson's staff.

Although Thompson's investments of men and money in Texas Gulf could only be regarded as one of his more profitable actions, it should have occurred to him that this rather wholesale transfer of talent to Texas Gulf might, in the long run, cost more than it was worth. At any rate, this exodus to TGS was the first and last such occurrence in the Thompson and, later, the Newmont histories.

William B. Thompson's successes with Nevada Consolidated, Nipissing, Inspiration, and Texas Gulf might have been enough of an accomplishment for any ordinary man's lifetime, but Thompson was in no sense "ordinary." Instead of being content to observe and to preside over the growth of one or more of these properties, Thompson launched into a series of promotions and stock transactions that rapidly developed a bewildering complexity, requiring more assistance and supervision than he alone could provide.

To handle his stock trades, Thompson founded his own brokerage house, and to give greater scope to his promotional activities, he established his own company, the predecessor of Newmont, called the Newmont Company, incorporated in Maine in 1916.

For the former, Thompson invited the man who became his brother-in-law, Walter Filor, to come down from Canada to join with two friends, Percy Bullard and David W. Smyth, to form the firm of Filor, Bullard & Smyth. Although Thompson kept the new firm fully occupied for years with his own transactions, the company eventually expanded to include general accounts and now operates a brokerage business quite independently of Newmont.

As his personal fortune grew, Thompson invested in oil and steel companies, other mining companies, and all sorts of ventures that caught his interest from time to time. A complete listing of Thompson's investments covered everything from Knox hats to Indian motorcycles and occupies several pages in the old account books of that period.

When Thompson incorporated the Newmont Company in 1916, he had apparently thought no further than the establishment of a vehicle for handling his larger acquisitions and promotions more efficiently. With a capital of only $500,000, the purpose of the new company seems to have been to trade in and to hold securities of other companies, inasmuch as not until the fourth paragraph of the Certificate of Incorporation is there any mention of conducting mining operations. Furthermore, if Thompson had any thought of selling stock in the Newmont Company to the public, there is no record of it, and not until 1925, after Newmont Mining Corporation, itself, was formed, did the company go public.

Newmont Company seems to have served Thompson's purposes admirably during the years of the First World War, and it was during those years that certain events took place that greatly affected the future of Newmont and of Thompson himself.

The first of these was his decision to energize Magma Copper Company, which owned a copper-silver mine near Superior, Arizona, that Thompson had bought in 1910 for $130,000, a precipitate action that shocked Henry Krumb, who had only recommended further study. When the First World War in Europe touched off an increased demand for copper, Thompson recalled Magma as a good bet, and in 1914 began active development of the mine. He had a power line brought in from Inspiration and started sinking a new shaft, which at the 1,000-foot level struck a high-grade vein of bornite, thus validating an earlier suspicion of Henry Krumb's. Along with building a mill, Thompson satisfied a lifelong ambition by ordering construction of a railroad to connect the Magma mine at Superior with the main line of the Southern Pacific. He could then talk on an equal footing (albeit facetiously) as a railroad president with his friend Tom Schumacher, the president of Western Pacific.

Incidentally, the Magma Arizona Railroad, with a total track length of 22 miles, is still running. It is reported that the only time the railroad made money was in 1963 when it was rented to a Hollywood crew to supply local color for the film *How the West Was Won.*

By activating Magma in 1914, Thompson unwittingly started an operation which became, fifty-five years later, a wholly owned Newmont subsidiary that is the largest contributor to the parent company's earnings. Surely one of the most fortuitous decisions Thompson ever made.

As the war progressed and the need for copper grew, investors on Wall Street were clamoring for copper stocks in a demand that seemed insatiable. Thompson had, in addition to Magma, an interest in an extremely high-grade but somewhat limited copper orebody in Alaska being mined by the Kennicott Mines Company of which Stephen Birch was president. Birch had brought this property to J. P. Morgan & Co. in 1907, which referred the prospect to W. B. Thompson who in turn approached the Guggenheims. Henry Krumb examined the property for them, and after seeing development headings in solid chalcocite, reported most enthusiastically on the richness and potential

productivity of the orebody. The Guggenheims organized the
Kennicott Mines Company to mine it and to finance the heavy
expense involved in building a railroad and a port in order to
ship the ore down to A.S. & R. at Tacoma. For years, these shipments ran between 60 and 70 per cent copper and around 15 oz.
silver.

Production by Kennicott proceeded on a fairly small scale
until June, 1915, when Birch asked Henry Krumb to go back
and reexamine the orebody to see if it would justify increased
production. Krumb concluded that the orebody was large enough
to justify an increase, but he suggested to Birch that the high
profits that would be produced at the higher level might more
prudently be invested in other copper properties than be distributed entirely as dividends. Birch thought well of the idea
and broached it to his principals, the Guggenheims. They in
turn called in Colonel Thompson, and under his direction there
occurred a series of exchanges of stock and stock purchases that
led in 1916 to the establishment of Kennecott Copper Corporation as the owner of the Alaskan mine, the Guggenheims'
Utah copper property, Nevada Consolidated, the Ray mine, and
the Braden copper property in Chile. The change to an "e" in
the company name was a clerical error. Within a month after
this consolidation, the common stock of the new Kennecott
Copper Corporation had a market value of $195 million.

These early war years found Thompson at his enthusiastic
best. Magma, Texas Gulf, and Kennecott all came at once, yet
Thompson found time to investigate and to try to promote still
another copper property in Canada. This discovery, which led
eventually to the formation of Hudson Bay Mining and Smelting Company, was made in 1914 by a prospector named Tom
Creighton, who was wintered in on a trap line near Phantom
Lake in northern Manitoba. Accounts of this discovery differ, but
certainly the more interesting of them is given by Arnold Hoffman in his book *Free Gold: The Story of Canadian Mining*. According to Hoffman, Creighton, in following a moose track,
came across an outcrop exposed by the wind in the heavy winter

snow. Creighton forgot the moose and chipped off a few specimens which later turned out to be full of chalcopyrite. After the spring breakup, Creighton came back with five partners and staked claims covering what appeared to be a large exposure of sulfide-bearing rock which extended out under the lake.

When it came time to choose a name for the entire property, the partners were prepared. Along their trap lines, they had come across a deserted cabin in which they had found a paperback novel called *The Sunless City*. The tale told the adventures of one Flinotin Flonneroy, who had accidentally wandered into an underground city whose citizens had amassed huge quantities of gold without ever realizing the value of it in the world above. The partners never found out what happened to the hero, because the former inhabitants of the cabin had torn out the last couple of chapters, probably for a more utilitarian purpose. Because Tom Creighton had panned some gold from oxidized rock on the shore of the lake, he announced that quite likely "old Flin Flon" did make it back to the surface of the earth and left this gold at the point where Creighton found it.

A syndicate formed to develop the Flin Flon property went to Hayden, Stone in New York for the financing. Becoming enthusiastic, Hayden, Stone agreed to spend up to $50,000 for drilling, but a combination of discouraging factors soon developed, and Hayden, Stone lost interest and withdrew. This initial program did bring in, however, Scott Turner, a former associate of Herbert Hoover who later became widely known as the director of the U. S. Bureau of Mines. In search of further financing, Turner was directed to William Boyce Thompson, principally because of the successful promotions of Nipissing, Inspiration, and Nevada Con.

Thompson looked over the maps and the reports and decided to send Henry Krumb up to Manitoba to look over the Flin Flon. Krumb wired back what was for him a wildly enthusiastic opinion expressed by the one word "excellent." Accordingly, Thompson had Krumb start exploring at Flin Flon in a campaign that ran through about $350,000 without, however,

finding ore enough to justify immediate production. Nevertheless, Thompson told Lamont about it, but by this time the situation in copper had changed from optimism to concern about
copper's position in the postwar world, and Thompson could
convince neither Lamont nor himself that Flin Flon deserved
then the large investment that would be required. Flin Flon's
day was coming, but was still years in the future. (See Chapter 3.)

Shortly after America entered the war, Thompson found
himself involved in a promotion of a different kind, one that
proved expensive financially and, more importantly, a serious
drain on his physical resources. Through a friendship with Herbert Hoover, Thompson had been a leader in encouraging mining industry executives to contribute to Hoover's Belgian War
Relief Campaign. Thompson's interest in this activity so impressed the executives of the American Red Cross that, when
they decided to organize a mission to Russia in 1917, they invited Thompson to be its head. Unpredictable as ever, Thompson accepted, and, leaving his affairs in the hands of Charles
Ayer, Henry Dodge, and Frank Holmes, Thompson, equipped
with the honorary title of colonel, went off to Petrograd, heading a staff with Raymond Robins second in command, a brilliant though radical young man also with a Western mining
background who later became Thompson's fast friend.

In Russia, Thompson found a new and absorbing interest.
Somewhat to the consternation of his old friends, Thompson became convinced that in order to keep Russia in the war and thus
occupy Germany on two battle fronts, it was essential that the Al-
lies recognize and support the Kerensky government. Thompson
put all his persuasive powers and a million dollars of his own
money into what turned out to be an unsuccessful effort.

Facing this fact, Thompson's clear vision and realistic brain
then told him that Kerensky could not survive, and that the Bolshevik group had the strength and the popular backing to win
out. He then tried to form ties between the Soviet government
and the Western world that might have served as a foundation
for mutual understanding that could have carried into the post-

war years. Here, again, Thompson was unsuccessful and was bitterly disappointed when the British and the Americans entered into, what was to him, an obviously futile military effort to overthrow the Bolshevik regime.

Thompson returned to the United States physically exhausted and somewhat embittered. Certainly his Russian experience left him considerably wiser in the ways of the world and in the cynicism of governments. The strain of that period of frustration in Russia unquestionably depleted his physical resources, and he was never again to be able to summon quite the energy and enthusiasm that had been his trademark.

Curiously, although he saw clearly and forecast accurately the course of the Russian revolution, he was unable, after he returned to the United States, to evaluate correctly his own potential effectiveness in American politics. Possibly the rebuff given him by the Montana politicians many years previously may have combined with his frustration during the Russian experience to cause him to want to succeed in politics as well as in finance. At any rate, in the early 1920's, Thompson once more embarked on an exhausting and, in the end, equally futile effort to accomplish a political objective. As a fund-raiser and as a contributor, he was quite acceptable to the Republican party hacks, but as a political officeholder, he was less than impressive. President Harding is said to have considered for him, but never offered, the Cabinet post of Secretary of the Treasury. In the end, Harding watered his gratitude down to the proffer of membership on a mission to a centennial celebration in Peru.

Even if he had assumed some elective or appointive post in a Republican administration, he would probably have been unable to serve; by the middle twenties, the trememdous burdens Thompson had laid on his own constitution began to show their effects. Throughout his life Thompson ate, smoked, and drank as he chose, though never seriously to excess. Without apparent effort, he could sit up all night and work all the next day, whether the all-night session involved a poker game or a frantic attempt to consummate a business deal. Thompson and his friends kept an apartment in the Murray Hill district of New York for no other purpose than to provide a place for their poker sessions. Calling themselves the "Sons of Hope," the group was quite capable of a continuous game for a day and a night or even longer. Relaxation with his friends was one of Thompson's chief pleasures and probably one of the last that he was willing to give up at his doctor's request.

This complex, rarely well-understood man yet had the power to draw to himself men of unusual ability in the various techniques he needed for his purposes. He had a host of friends and he enjoyed hugely his poker sessions and his mining promotions that did bring him a personal fortune. Yet underneath it all, he seems never to have achieved lasting happiness for himself and his family.

To an extraordinary degree, Thompson typified the quality mentioned in Kipling's "If," in that he could "walk with Kings —nor lose the common touch." Whether meeting with statesmen, financial giants, engineers, or workmen, Thompson was equally at home. Yet he was forever uneasy and unfulfilled. In the end, only in the garden he had created at his home in Yonkers, New York, or in the arboretum he established at his Picket Post House at Superior, Arizona, could he really relax and let his spirit expand.

As the years went by, Thompson's activities were more and more restricted by failing health until his life became a succession of trips to Florida, back to Yonkers, a cruise on his yacht, the *Alder,* and a stay at Picket Post House. It is clear that the years of furious activity before and during the war had drawn too heavily on Thompson's strength. His health declined rapidly in the late 1920's and on June 27, 1930, at the age of 61, he passed away at his home in Yonkers.

At a meeting following his death, the board of directors of Newmont adopted a resolution saying in part, "No man did more to place mining enterprises, and more particularly, the business of copper mining, within the pale of legitimate effort and of industrial investment.

"By application of his precept and example, and from the inspiration of his character and memory, these directors, who were his friends, can make of this corporation an adequate realization of his vision and an enduring monument to his genius."

Those who have followed in Newmont have done just that.

Fred Searls' birthplace, Nevada City, California

2 A New Team Takes Over . . .

When William Boyce Thompson replaced the old Newmont Company with Newmont Corporation in 1921, his intention appears to have been to establish a larger and wider-ranging company that he would handle in the same energetic and highly personal manner that had been so successful in the past. However, exhausted from his Russian experience and having built a fortune that at his death was estimated at about $150 million, Thompson no longer felt quite the old urgency for sheer money-making. Almost at once and in various ways, some of them subtle, and some quite obvious, Newmont began to change from the family-owned-and-directed enterprise Thompson had conceived, to a company influenced in its day-to-day operations by others. By the end of the decade of the twenties, Newmont had become a widely known and respected publicly owned company. It was not, however, a well-understood company, even by some of its own administrators. In developing out of the Thompson era, Newmont experienced some growing pains, but it also attracted to its management as skilled and as individualistic a group as any mining company has ever had.

Incorporated in Delaware in 1921, Newmont Corporation was capitalized at $8 million expressed in 500,000 shares of common stock and 300,000 shares of preferred, both of $10.00 par value. Most of these shares were held by the Thompson family, although a few shares were distributed among the officers, board members, and Thompson's friends. The principal asset of Newmont Corporation was, of course, Thompson's portfolio of mining and oil stocks.

In its first two or three years, Newmont did little to distinguish it from the holding company it may have appeared to be. Always restless and impatient, Thompson became dissatisfied. His remarkably acute sense of timing told him that to continue to grow in the mining industry, his new company could not remain purely a holding company. Clear evidence of his appreciation of the changing climate in which Newmont was to operate exists in the form of a handwritten note still in the Newmont files that Thompson wrote in 1925.

This note reads, "Make mining company. Not converting stocks into bonds or cash for fear might be spent too fast or unwisely. Leaving in *good* mining company stocks which can sell when need money for a better venture."

Although there is nothing to identify the precise purpose for Thompson's having written this note, the best guess is that the scrawl was intended as a reminder for use in connection with a statement he intended to make at a board meeting. At any rate, his intention and the result are clear, for in 1925, the name of the company was changed to Newmont Mining Corporation. At the same time, 130,000 of the company's 430,000 common shares outstanding were sold to the public at an initial price of $40.00 per share. Trading in Newmont began on the New York Curb Exchange (now the American Stock Exchange), Thompson's favorite market. If one had bought 100 shares of Newmont at $40.00 in 1925, and neither bought nor sold thereafter, he would now own, through stock splits and dividends, 4,330 shares, worth, as of February, 1972, about $140,000, which would yield $4,503.20 per year in dividends.

Thompson's note should really have been framed, because the philosophy expressed there has guided the company ever since, with the notable exception of the period 1928–32. Under the three chief executives who followed Thompson, only once has there been any doubt within the company that Newmont Mining Corporation was, indeed, engaged only in the business of exploring for, acquiring, financing, developing, equipping, and operating mining and petroleum properties.

In 1925, however, the precise nature of the Newmont enterprise was not a matter of public concern. All the public knew, or cared to know, was that Colonel Thompson had an excellent reputation as a wise and shrewd investor, and that Newmont itself reflected this reputation. Even though by 1925 Thompson was content to remain as chairman and leave the major responsibility of directing the company more and more in the hands of its second president, Charles F. Ayer, the financial community was still thinking of Newmont in terms of Thompson's leadership.

Nevertheless, it was Ayer's influence that was growing as Thompson's declined. Ever since 1907, when Ayer came into Thompson's organization as a vacation replacement for Knobloch, Thompson had steadily gained in appreciation of Ayer's legal ability. Thanks to Ayer's skill, Thompson had never had to spend a day in court over mining claim litigation. There were disputes now and then, but owing to Ayer's expertise as a negotiator, settlements were always reached out of court.

Ayer's one quirk was a temper that occasionally mastered him. Meticulous in most things, always well groomed, usually reserved and polite, it was the more startling when a fit of anger burst to the surface and broke over the head of whoever had inspired it. Even more shocking was Ayer's use in these outbursts of certain choice Anglo-Saxon expletives. Such displays, however, were rare.

All his life, Ayer kept a daily record, not really detailed enough to be called a diary. From this and from his correspondence files, one gains the impression of a competent, careful man of unswerving integrity, but of a much less imaginative and dashing turn of mind than was Thompson. Where his predecessor would have ignored or overridden strong opposition within the company, Ayer tended to listen to all points of view and to attempt to achieve an accommodation among them. Recognizing his limitations, Ayer conceived his position to be that of a mediator among the several strong and fiercely individual personalities who were drawn to Newmont in the late twenties and early thirties, a concept that appears actually to have benefited Newmont under the circumstances.

It is also true that from the very first, Ayer was backed up by an unusually competent and interested board of directors. When Newmont went public in 1925, the board included Thompson as chairman and, as outside members, his friends Albert Wiggin, president of the Chase Bank, John Kemmerer, a coal-mining company owner and executive, Stephen Birch, president of Kennecott Copper, and Henry Krumb, his consulting geologist. Vernon Munroe represented Tom Lamont and the Morgan bank. Inside directors included Alexander J. McNab, David E. Thomas, Henry E. Dodge, and Fred Searls, Jr., Searls having joined Newmont in 1925.

These men, as do Newmont board members today, took their positions seriously and as a result felt themselves very much a part of company affairs. The philosophy underlying the selection of Newmont directors appears in a letter Judge Ayer wrote to Margaret, W. B. Thompson's daughter, in November, 1930. This letter read in part:

When your father's health broke, there was public concern as to what would happen if he should suddenly pass away, and he let it be known that he had built up an organization and had arranged for it to carry on in his absence. The public always looks, of course, at the list of the board of directors of a company to see who is going to have the investing of the funds. Your father wanted a strong board for Newmont, desiring to present in its membership what he called a good picture to the investing public. A director in a live, growing company, as Newmont is, should give careful thought and constant attention to the direction and management of its affairs. I am sure that your mother feels that the present directors are looking after the interests of the corporation to the best of their ability. None of them regards his position as merely an honorary one.

However, a strong board of directors can be of even greater value with a strong leader within the company. Newmont acquired a leader in 1925 in the person of Fred Searls, Jr. Because of his growing awareness that Newmont would have to do more mining and less promotion, Thompson correctly concluded that the real future of Newmont lay in actual exploration for mineral properties, an activity that, despite his many successful promotions, Thompson had not pursued as vigorously as he might have. In Fred Searls, Thompson found a man whose experience and capabilities matched the company's needs, and the Searls era in Newmont was under way.

Newmont has certainly had its share of strong and colorful personalities, and Fred Searls is outstanding among them. Stories about him are endless, but all are entertaining. People who knew him well have characterized him as being generous, stingy, courageous, fearful, uncommunicative, forthright, decisive, hesitant, intuitive, and pragmatic. He could have been described in all these terms, because the fact is that at various times he was all these things and more. All who knew him agree, however, that Searls was an excellent exploration geologist, one of the best, especially in the early years with Newmont, and that he did supply the company with leadership. Because of his strong influence on the company, it is well worth examining Fred Searls's background.

Niles Searls, Fred's grandfather, left Rensselaerville, New York, in early 1849 headed for California. He arrived there in August, 1849, stopping at what is now Marysville. Soon giving up the idea of making his fortune in gold mining, Niles Searls turned to using his legal training to help settle litigation over conflicting claim ownership that rapidly sprang up in the gold camps.

Searls eventually returned to New York, married his first cousin, and the couple returned to California by way of the Isthmus of Panama. In California, the Searls settled in Nevada City, where they built a home that is still standing and still occupied. Niles eventually became Chief Justice of the California Supreme Court.

Niles's son, Fred, Sr., also followed the law, maintaining his headquarters in Nevada City. Fred, Sr. married Helen Pond, who was the daughter of a Congregational minister. She taught her six children, five boys and one girl, for the first two or three years of their schooling, basing much of her instruction on the Bible. This teaching stuck with Fred Jr. for years, and he used the technique of parables to explain complicated geologic points years later whenever he appeared as an expert witness in mining claim litigation.

Fred Searls grew up in Nevada City, surrounded by a mining atmosphere and in ideal climatic conditions for an outdoor life. He and other boys in the neighborhood belonged to an association resembling the Boy Scouts which was called "Coming Men of America," or CMA for short. This outdoor tradition and the CMA also stuck with Fred all his life. For example, when bedtime arrived on the rare occasions when Fred was home from his field trips, his wife would announce it by giving Fred the CMA salute.

Fred went to the University of California at Berkeley to study geology, graduating in 1908. He stayed an extra year to teach mineralogy and to act as assistant to Professor Andrew Lawson, a geologist who did a great deal of consulting work in apex litigation.

After leaving the university, Fred Searls established an office in San Francisco with a friend of his named Joe Thorne. From this office, the two did consulting work as geologists all through the West. Searls made friends easily and he had a host of them. It was characteristic of him throughout his working life that whenever he needed outside expert advice, he would draw on the many professional friends he knew and trusted.

In his travels to various mining districts, Fred Searls met Seeley Mudd and, through him, met Bernard Baruch and William Boyce Thompson. Early in his career, Searls did some mine examination work for Thompson up in Montana, and although nothing productive resulted, Searls's performance impressed Thompson sufficiently so that when Thompson at one point set up a company to explore certain areas in China, he asked Fred Searls to examine the properties for him. Europe was at war when Searls went to China, and when the United States entered the war in 1917 and the news reached him, he immediately returned to the United States to enlist in the army.

The story goes that when Fred got back to California, he called up two friends of his who were working near Grass Valley, Billy Simkins and A. F. Duggleby. He asked both of them to meet him the next day at the courthouse in Grass Valley. When they met Searls, they asked him what was up. "Boys," he said, "we're going to enlist." Possibly it took more argument than this simple statement to convince the two friends, but Searls was nothing if not persuasive, and enlist they did. He and Simkins and Duggleby were able to stick together throughout their military service. Enlisting as a private, Searls was promoted in the field to sergeant and later to lieutenant.

After the war, Searls went back to California and took up his consulting work once more. In early 1921, he went back to China for one of the Thompson enterprises. Fred's assignment was to visit certain mines in southwestern China, at least one of which, the MuNai, was said to be as rich as the Bawdwin lead mine in Burma.

Typical of Searls's concentration on the job at hand was his mention in his final report on the trip that "we arrived in Puerh on the 22nd having been delayed by bandits." This is his only reference to bandits, although there exists a story that Fred was

William Boyce Thompson, founder of Newmont Mining Corporation, in the garden of the mansion he built in Yonkers, N.Y., in 1909. Only forty, he was already several times a millionaire.

Charles F. Ayer, Newmont's president from 1923 to 1946 and its chairman from 1946 to 1954. Affectionately known as "Judge," he maintained the highest standards of integrity and competence in his legal work.

Fred Searls, Jr., won world-wide fame and friendship as a top-notch exploration geologist and mining executive. He was Newmont's president from 1946 to 1954, and chairman from 1954 to 1966. *Roy Stevens photo*

Henry Dewitt Smith, one of the mining industry's great engineers, joined Newmont in 1929. He was especially well-known for guiding development of O'okiep and of Tsumeb in southern Africa.

Franz Schneider joined Newmont in 1930 as a vice president and is still (1972) active as a consultant. His financial genius powerfully aided Newmont during the Depression and in the succeeding years of growth.

George A. Kervin, an all-around "good man," helped McNab at Mason Valley in 1917, explored O'okiep in the late 1920's, got O'okiep going in 1937-41, and died following an accident at the Gray Eagle mine in 1941.

A. J. McNab, whose determination brought the San Manuel copper project to fruition, joined Magma Copper Company in 1911, and served Magma and Newmont capably and loyally the rest of his life.

actually captured by bandits and had to be ransomed. If true, he did not consider this worthy of mention in his geologic report.

At any rate, Searls and two other geologists, accompanied by porters carrying their supplies and Chinese soldiers to guard them against hostile natives, reached the MuNai mine. However, as soon as they had arrived, two Kawa tribesmen delivered an ultimatum in the form of a bamboo stick notched on the edges. Tied to the stick were a red pepper, a piece of charcoal, and some feathers. The Chinese soldiers, when shown this item, said that this meant that the nearby Kawas were very angry about this intrusion. The feathers meant that the Kawas would come as swift as a bird can fly. The charcoal meant a burned camp, and the red pepper, a murdered party. Having said this, the soldiers immediately packed up and left. The geologists stayed long enough to go through the mine, but left the next morning, just ahead of the Kawas.

The area visited by Searls had attracted attention because of the presence of large slag heaps which were the residue of the primitive lead smelters operated by the Chinese, principally to recover silver. It was assumed that the slag heaps had been associated with lead-silver orebodies. Searls's investigation led him to conclude that the slag heaps were derived in large part from the smelting of lead-silver ore fragments that the Chinese had taken from alluvial deposits in sinkholes, of which there were a great many throughout the area.

Fred reported that these sinkhole ore concentrations were new in his experience, although sinkholes themselves are typical of limestone country. Because of what he regarded as a discovery, original with himself, he studied the sinkhole concentrations quite thoroughly and proved to his satisfaction that there was very little ore in place anywhere in the area.

It is interesting to note that Fred searched the literature when he returned to the United States in order to check his belief that the sinkholes he had seen had not previously been reported elsewhere. He found that similar concentrations of lead and zinc in sinkholes did occur in Dalmatia and in the Tri-State

District in the United States. Actually, during the Korean War, because of emergency needs for lead and zinc, many such sink-hole deposits were mined in the Tri-State District after the major orebodies were worked out.

Shortly after his return from China, Searls took a job with U.S. Smelting, Refining and Mining Company and moved his family from Nevada City to Salt Lake City, Utah. From Salt Lake, he continued his work in field geology, but because of Thompson's high opinion of him, he eventually received an offer transmitted to him by Judge Ayer. Hearing of this, U.S. Smelting made Fred a counteroffer. Following an exchange of telegrams between himself and Thompson, Searls agreed in June, 1925, to come with Newmont at a salary reported to be $40,000, which Searls specified must be considered on a ten-month basis because he said he wanted to spend some time with his family. His son, Bob Searls, comments on this, "We never saw him anyway."

Searls came east alone and after a preliminary discussion with Judge Ayer, asked Ayer to recommend a good place to live in the New York area. Ayer said that most of the Newmont group lived in New Rochelle, and it seemed like a good place. Fred asked if Ayer knew of any houses available in New Rochelle, and Ayer said there was one about a block from his own home. Fred replied, "I'll buy it," which he did. Bob Searls reports that his mother never really forgave her husband for this hasty action in buying a house, sight unseen.

Few men in the mining industry have so quickly become a legend in their own time as did Fred Searls. He rapidly acquired a reputation throughout the industry for an ability to assemble quickly in his mind all the relevant facts that he could obtain and reach a decision in a lightninglike process that many people found difficult to follow. To some of his associates, this process seemed to resemble a response to a hunch rather than to a logical process of thought. Indeed, there exists until today a good deal of difference of opinion as to how much of Searls's thinking was based on intuition rather than the careful consideration of facts.

Also legendary was Searls's appearance. Early in his career, Fred began affecting a certain eccentricity of dress that became his trademark. From his old cloth cap to his yellow mining boots, his appearance was extraordinary. He always wore a red "jazzbow" necktie, nondescript jacket, and trousers that were so short as to show white socks above his shoes. In the beginning, this affectation probably arose from his realization that this sort of dress tended to relax the opposition when he appeared in court. Opposing attorneys in mining claim litigation, on seeing Fred in his customary outfit, were probably prone to believe that in Searls they had somewhat of a country bumpkin with whom to deal. This natural mistake was usually their undoing.

For example, Henry DeWitt Smith once told a story about Fred Searls's appearance in a court in Idaho in a case involving apex litigation. Appearing for the plaintiff in the case was a distinguished Idaho geologist. When this geologist took the stand, the attorney reviewed his qualifications, which included ten years' work as a consultant and as a geologist with the U.S. Geological Survey. The attorney asked the geologist to what societies he belonged, and the man mentioned membership in the Geological Society of America, The American Association for the Advancement of Science, The Royal Geographic Society, the Society of Economic Geologists, The American Institute of Mining Engineers, and so on. It appeared that the jury was much impressed, for they listened solemnly to his abstruse testimony on ore deposition.

Then Fred Searls was called to the stand to appear for the defendant. Fred tucked his old cap into his pocket, adjusted his frayed bowtie, and walked to the stand, his white socks showing above his yellow boots. "Mr. Searls," asked the defendant's attorney, "of what societies are you a member?" Fred scratched his head, thought a minute, then answered "Well, I belong to the Elks." This broke up the courtroom, and following Fred's testimony, the jury gave its verdict to the defendant.

This was the sort of man who, in 1925, went actively to work, looking for mineral properties quite literally all over the world. In his restless examination of prospects, he covered practically all of Africa, China, Turkey, Persia (Iran), Mexico, Peru, some old Roman gold mines in Spain, as well as hundreds of prospects in the United States and Canada. Fairly soon it became apparent even to Fred that there was just too much work for one man to do successfully.

In typical Searls fashion, he turned for help first to his old friends and comrades in arms back in California. He put Billy Simkins to work on examinations in California. Duggleby he sent on foreign explorations, in which he was engaged for Newmont for many years. One such trip took Duggleby to a property in South West Africa, which Duggleby thought was quite interesting but which could not be pried loose from its German owners. That was Newmont's first look at the mining center of Tsumeb.

Duggleby met a tragic end in the Philippines during World War II. At the time of Pearl Harbor, Duggleby was working for Benguet Mining Company and was captured by the Japanese during the invasion. He was imprisoned at Santo Tomas in Manila. A fellow prisoner was accused of a minor infraction, and because of this man's poor physical condition, Duggleby took the blame. The Japanese marshaled the prisoners in the courtyard, brought Duggleby out before them, and executed him on the spot.

Another of Fred Searls's friends from California, George Scarfe, was active for Newmont in Peru. Searls also brought in P. C. Benedict whom he had met years previously at the United Verde property in Arizona. Benedict joined Newmont in the late 1920's, and made the first of many outstanding contributions to Newmont at the Island Mountain mine in British Columbia in 1934.

Most significant of the additions to Newmont Fred Searls made in his efforts to build an exploration staff was Philip Kraft. A native New Yorker, Kraft was a graduate of Columbia University as a mining engineer and geologist, reportedly because his father insisted on it. Kraft completed his mining engineering studies in Germany, where he met the American girl who became his wife. She was also at that time studying in Berlin.

Kraft's first position in the mining industry was with Dome Mines Limited in Canada. He worked as a geologist for a brief period, following which he went to California for the Bishop Oil Company, one of the properties of William Wallace Mein. Working in California, Kraft became well acquainted with Henry Krumb, who recommended him to Fred Searls as a competent geologist.

Because of his experience in Canada, Kraft was able rapidly to take over Newmont's Canadian exploration activities, and he was later responsible for several of the producing properties in the North Country. His chief activity, however, eventually came to be the management of Newmont Oil Company, although he never quite abandoned his interest in Canadian exploration.

In addition to his ability as a geologist, Kraft had an extraordinarily keen intelligence and a wide range of interests both inside and outside his field of technical competence. Acquainted in the areas of art, literature, and music, he also displayed a sympathetic awareness of the problems and the ambitions of other people. Steadfastly loyal to his profession, to his friends, and to his company, Kraft's quiet influence on Newmont was remarkably strong for one so little inclined to push himself forward. Mention of Phil Kraft's name immediately calls to mind his earnest countenance framed by spectacles dangling from one ear or pushed back on his forehead, as he discussed with dry wit and cogent argument the point at issue.

Phil Kraft was never satisfied. Intellectually, he was a perfectionist, seeking always that most elusive of objectives, the one "right" action. Abhoring argument and dissension, as he did, he yet defended his positions stoutly, yielding sometimes only to the greater weight of the Malozemoff logic or the Searls determination.

Phil Kraft had a long and productive, although often frustrating, career with Newmont. He continued working through his retirement age and died in December, 1968, not long after he

had seen launched a dream of his, which was the successful as-
sembly of a French and American exploration consortium that
found oil in the Algerian Sahara Desert.

Along with these relative newcomers, there were in Newmont in 1925 three of the stalwarts from Thompson's original
staff: Charles F. Ayer, A. J. McNab, and Henry E. Dodge. Ayer,
who has already been described, was president, Henry Dodge
was secretary and treasurer, and McNab was then a Magma vice
president and a Newmont director.

Born in Ontario, McNab obtained both a Bachelor of Arts
and a Bachelor of Science degree from Queen's University. After
graduation, he went to work in the metallurgical department of
Consolidated Mining and Smelting Company at Trail, British
Columbia. His metallurgical talent soon brought him promotion, and McNab was given the responsibility for the Betts electrolytic lead refinery at Trail.

After W. B. Thompson got McNab away from Consolidated
Mining and Smelting in 1911, he made use of McNab's particu-
lar skill in designing and building the copper smelter for Mason
Valley Mines Company, which began operating in 1912. (See
Chapter I.)

When a drop in the copper price in 1919 forced Mason Valley to shut down, McNab came to New York, where he fitted
quite naturally into the management of Magma Copper Company. When Newmont was formed in 1921, it shared office space
with Magma, as it has ever since. McNab became a vice president of Magma in 1920 and a director in 1923. Although he also
served as a director of Newmont until 1926, and again from
1931 to 1948, McNab devoted his chief energies and his unflag-
ging faith to Magma.

A slight, slim man with thinning hair, piercing eyes, and
beetling eyebrows, "Mac," as he was known to his Newmont associates, or "Sandy" as his old Canadian friends called him, was a
nervous, restless individual, strung tight as a coiled spring. His
staccato, laconic delivery of his opinions, sometimes expressed in
barely formed sentences, belied an unusual clarity of mind and
an almost infallible judgment. Had this faculty been less sound,

he would have met endless trouble, for his will and determination were of iron. Seemingly living on a short fuse, McNab was yet a friendly, modest, and solicitous man. Although he could explode in anger and derision over half-baked opinions, inadequate knowledge, or faulty judgments, he was respected by everyone who knew him, even those who had felt his displeasure.

In contrast to McNab, Henry E. Dodge was quiet in speech, and if he ever lost his temper, none of his associates remembers it. Despite his unassuming manner, he was nevertheless one of the strongest influences behind the progress of both Newmont and Magma. Although he had a natural talent for finance, largely self-developed, he resembled in no way the common conception of a cold, stony-hearted financial genius.

Dodge was one of the few men foresighted enough in 1929 to sell his common stock holdings at near the peak of the market. After the Crash, instead of crowing over the plight of those of his Newmont associates who had ignored his advice, he covered several of their margin accounts with his own funds to prevent their being entirely sold out, certainly an act of friendship, not of cold-blooded calculation.

A native of Nova Scotia, Henry Dodge, as a young man learned railroad telegraphy, perhaps because of the opportunity thus offered for travel. He got a job as telegrapher with the Canadian Pacific Railroad and was moved before long to the station in Winnipeg. One afternoon a message came through from Consolidated Mining and Smelting Company at Trail, British Columbia, which was affiliated with the CPR, mentioning an opening for a bright young man in the Trail business office. Dodge forwarded the message to Toronto, but he also wired Trail offering himself.

Impressed by Dodge's initiative, the Trail office accepted him, and Dodge rapidly acquired a good working knowledge of the accounting and financial aspects of mining company operations. In fact, Dodge's performance became so outstanding that Walter Aldridge, Cominco's chief engineer, became aware of his ability and remembered it when he left Cominco to join W. B.

Thompson in January, 1911. After a short time, Aldridge decided Thompson really needed another man to run his office and recommended Dodge. The result was that on Election Day in 1912 Henry Dodge began working for Thompson, who promptly sent him out to Arizona to help with the newly formed Magma Copper Company. Dodge was a director of Magma from 1913 until his death in 1954.

Henry Dodge's name appears in the records of most of Thompson's, and later Newmont's, affiliates. He was treasurer or secretary of Hudson Bay, Mason Valley, Magma, Newmont, and several others. In the early days, however, he was busy simply recording Thompson's active stock market trading, which kept him in the office night after night to have the data ready for the colonel the next morning. Taller than either Ayer or McNab, and of fair, rather delicate complexion, Dodge's appearance matched his temperament. A listener rather than an arguer, Dodge made his points by a simple recital of the facts, which he usually knew more thoroughly than anyone else. Few men have been as loved and respected by the men who worked with him and for him as was Henry Dodge.

This completes the description of the emergence in Newmont Mining Corporation of an administration composed of several strong individuals, in comparison with the essentially one-man operations of the Thompson era. This new administration proved itself capable of carrying out successful exploration, and then going on to finance worthwhile properties and assume responsibility for their development and their management as well. It would be pleasant to be able to report that the Newmont management thereafter held to this basic source of strength, but it would not be true. Diverted by speculative fever in the late twenties, Newmont's leaders nearly lost sight of their established purpose, and in the early thirties they ran into serious difficulty as a consequence. The actions that on the one hand endangered and on the other preserved Newmont make an absorbing chapter in the Newmont history.

Thompson's Touchstone, 1925

3 ... And Meets the Crash

Measured against such successes as Inspiration, Kennecott, and Texas Gulf Sulphur, the accomplishments in actual mining of the fledgling Newmont administration during its first decade were somewhat less than impressive. True, Newmont had much to do with launching Hudson Bay Mining and Smelting Company in 1928, but otherwise the new company got off to a rather slow start, as a mining company that is.

However, considered as an investment vehicle, the pace was a lot faster. Despite his expressed desire to form a true mining company, Thompson could no more resist the urge to trade in mining and oil shares than he could stop smoking his favorite cigars. Furthermore, his trading fever was as contagious as the mumps, and some of Thompson's associates inevitably came down with it. In the late 1920's they gradually turned more and more of their real interest away from mining toward the much more exciting and faster-paced activities in the stock market.

Therefore, an account of Newmont in its first decade must cover two general areas of activity. In the one, Newmont sought to explore for, invest in, and develop promising mineral properties, just as Thompson had intended. In the other area, Newmont invested in the stocks of other mining and oil companies, not entirely to establish an income-producing liquid fund for financing new ventures, but simply as a means of speculation. The latter activity brought Newmont perilously close to disaster, but the former activity, in a way no one could have foreseen, kept Newmont going. Before taking up Newmont, the speculator, it might be preferable, therefore, to consider Newmont, the miner.

First, a word about Magma Copper, the company the colonel had organized back in 1910. Although today Magma, as a wholly owned subsidiary of Newmont, stands in the front rank of U.S. copper producers, it was not, in the 1920's, highly regarded as a "Newmont" property. In fact, Newmont did not add Magma shares to its portfolio until 1928, and its ownership of Magma never exceeded 15 per cent until the first attempt at financing Magma's San Manuel project in 1949, when Newmont's holding rose to 22.9 per cent.

Yet from the very first days of Newmont's existence, Magma and Newmont shared headquarters office space and had officers and board members in common. To appreciate that the two companies could be, and actually were, operated quite independently of one another, one must understand something of the fierce independence of A. J. McNab, and also the absolute integrity of McNab, Judge Ayer, and Henry Dodge, all of whom served both companies. It was an unusual arrangement, but these were unusual men. Because for years these men resolutely kept Magma apart as a company on its own, it is only fair to record Magma's history quite as separately. It appears, therefore, in Chapters 8 and 9.

Newmont's attempts at mine development in the 1920's stemmed largely from the activities of Fred Searls, but before launching into these, one should consider a highly successful venture in which Fred took but little part. This was the activation, through Hudson Bay Mining and Smelting Company, of the property at Flin Flon, Manitoba, that Colonel Thompson had tried so hard to promote in 1919 and 1920.

When the colonel and Hayden, Stone gave up on Flin Flon, the group headed by Scott Turner and R. E. Phelan did not. They hung on during the lean years of the early 1920's, and in 1924 they approached the Harry Payne Whitney family with Flin Flon as a prospect. The Whitneys called on a mining engineer, Roscoe H. Channing, Jr., to evaluate the property. Needing geological help, Rock Channing, as everyone called him, asked Henry Krumb to visit Flin Flon. Krumb said he didn't need to, went over to his file and pulled out his original report on Flin Flon. On the strength of it, backed by Channing's own studies, the Whitneys took up the option on Flin Flon in 1925, thereby acquiring what turned out to be one of the great mines in North America.

When news of Whitney's action reached Judge Ayer, he sent McNab up to Manitoba for another look. Ayer reasoned that, inasmuch as Newmont had already spent around $350,000 on the property, Newmont should not overlook any possibility of getting at least some of its money back. This was among the most profitable moves that Ayer ever made.

McNab was quite as enthusiastic as Henry Krumb had been previously. "The property is situated in a new district and little prospecting has been done," he wrote. "Nevertheless, this is a large area of geologically favorable rock in a general way similar to the areas that have produced the great mines in Ontario. While the capital requirement is high, I do not know at this time where money can be placed in a mining enterprise that promises as much return with equal safety."

As it turned out, a good deal more than enthusiasm was required. Mining the orebody wasn't too difficult, but concentrating the copper and the zinc presented a real problem. Actually, the sulfide minerals composed more than 90 per cent of the ore, so that in effect, the copper, zinc, gold, and silver occurred in a pyrite gangue.

Following thousands of flotation tests made at the Colorado

School of Mines, a flow sheet for a pilot mill at Flin Flon was worked out in 1927, which involved separation of copper and zinc, followed by cyanidation of gold and silver in the flotation tailing. Results at first were quite discouraging, and as word of the poor results spread, the project was referred to by investors, for a brief period, as "Flim Flam." Even after successful test work, establishment of the flow sheet in the commercial mill at Flin Flon proved difficult. In the end, the problem was solved, not by some magic of technique, but by the development of operating skill through experience in the mill.

Eventually, results improved enough to justify the formation of Hudson Bay Mining and Smelting Company in 1928 to operate the Flin Flon property. Newmont went in for 350,000 of the 1.million shares issued at $10.00 per share. Newmont also offered its own shareholders rights to buy another 237,804 shares of Hudson Bay at $15.00 per share. Response was strong, and the price on the New York Curb quickly rose to over $20.00. Despite the now promising outlook, however, things were not too good when McNab visited Flin Flon again a few weeks after the company was formed. He found a considerable amount of confusion and dissension among the principals at the property. In a letter to Judge Ayer, he wrote: "If serious measures aren't taken to improve the situation, the success of the enterprise will be imperilled."

In response, Judge Ayer took action in a way that became characteristic of Newmont. Although Newmont had less than a majority interest in Hudson Bay, Ayer was able to persuade the other shareholders that Newmont should have a large share of responsibility for bringing the mine into production. Somewhat reluctantly, the others agreed, and from then on for many years the affairs of Hudson Bay were largely directed jointly by Rock Channing and McNab.

Quite similar in appearance, they were also similar in character and hit it off amazingly well. They shared strong qualities of integrity and perseverance, some called it stubbornness, but fortunately they thought alike on most corporate problems.

They became close friends, and McNab served on the Hudson Bay board to good effect, and in turn, Channing aided McNab as a director of Magma Copper Company from 1937 until 1958.

Hudson Bay entered production in October, 1930, hardly an auspicious year. Metal prices dropped, and the price of Hudson Bay stock dropped with them, sagging below a dollar in 1933. That was the worst year, however, and the determination and skill of Messrs. Channing and McNab began to pay off in the form of increasing production, increased metal recoveries, and lower costs, just as metal prices began firming up once more.

In the first fifteen years of its life, Hudson Bay produced metals worth nearly $220 million, and paid dividends of over $52 million, a performance that made the property one of the most highly regarded items in the Newmont portfolio. However, by the early 1950's, ore at Hudson Bay became a little harder to find. Fred Searls and Henry Smith began to wonder if, indeed, the orebody might not be bottoming out, and by 1953, they were almost sure of it. McNab reluctantly agreed, and Newmont sold out its Hudson Bay shares, the last of them going in 1954.

Thereafter, Newmont watched from the sidelines as Hudson Bay found more ore and went on to even greater profits and dividends. In 1970, Hudson Bay had a net income of $19.6 million as against $29.7 million in 1969, but in 1971 a long strike cut drastically into earnings and net income dropped to $3 million. Nevertheless, Hudson Bay's outlook continues strong.

Certainly, Searls and the others later regretted pulling out of Hudson Bay too soon, but they had no reason to be anything but proud of Newmont's part in Hudson Bay's early history. This episode had served notice on the mining industry of the 1930's that in Newmont there had developed a financial and technological power to be reckoned with and a competitor to be respected.

However, as important to Newmont as Magma and Hudson Bay eventually became, the story of Newmont, the miner, in the 1920's is essentially the story of Fred Searls's search for profitable mining properties, and this involved neither Magma nor Hudson Bay. One of Searls's first enthusiasms after joining Newmont in 1925 was the opportunity to take part in the activities in South Africa and Rhodesia of a South African company, Anglo American Corporation of South Africa, Limited, directed by Ernest Oppenheimer. Colonel Thompson had helped form this company in 1917, when Tom Lamont asked him to organize a syndicate to handle American participation in Oppenheimer's new company.

In forming the new company, the American syndicate, headed by Thompson, came in for 25 per cent of the initial capitalization of £1,000,000. In addition to Thompson, the syndicate included W. L. Honnold, who had initiated the idea, J. P. Morgan & Co., and the Lewisohn interests, as well as Herbert Hoover. Anglo American Corporation became even more powerful than Oppenheimer dreamed, and when Thompson incorporated Newmont, he added the participation in Anglo American to the Newmont portfolio.

Although Fred Searls and Ernest Oppenheimer are said to have met in 1913, the first real contact between the two men took place in 1925. Two men more unlike each other would be hard to imagine. Oppenheimer, although born in Germany and a naturalized South African citizen, was the very personification of the smooth, well-tailored, urbane Englishman. Searls was a tough, out-of-doors' man, careless in dress, almost to the point of a calculated carelessness, and not much given to the niceties of drawing-room conversation. Nevertheless, there developed a mutual respect between the two, and they kept close contact for years, although there never developed a really close cooperation.

In only one respect were Oppenheimer and Searls much alike. Both men could, in their thinking, drive so rapidly to the heart of the matter that the process appeared to some of their associates to be intuitive. Ernest Oppenheimer's son, Harry, once described his father as a man who frequently acted on the basis of a hunch, something a good many of Searls's friends also said of him.

For the purposes of the present chronicle, however, the important point is that Searls developed a continuing interest in the affairs of Anglo American. These were much too complicated to be reported in detail here, and adequate records are available elsewhere,[1] but in essence, a competition developed in the mid-1920's between Anglo American on the one hand, and Rhodesian Selection Trust on the other. Although the two companies were interrelated to some extent, Oppenheimer was basically afraid of RST because of the strong interest that American Metal Company had in the latter. Oppenheimer had great respect for A. Chester Beatty and American Metal, but he had no intention of letting them wield great influence in Anglo American. Beatty was an American mining engineer, prominent at first in American Metal Company, who later became a British citizen and helped organize Rhodesian Selection Trust in which American Metal developed an active interest.

To help form another new company that he hoped would become as strong as RST, Oppenheimer, in 1928, again asked Tom Lamont of J. P. Morgan & Co. for advice as to how matters should be handled, particularly with reference to securing additional American help, other than from American Metal Company. Lamont, in return, asked why Newmont had not been considered as a participant. Leslie Pollak, Oppenheimer's brother-in-law, replied that they had never thought of Newmont. "We had no direct dealings with them. We don't know their policy, aims, or scope of operations." This seems a strange statement, considering the close connection among Thompson, Searls, and Oppenheimer. It also foreshadowed further puzzling developments of later years, but at the time, Searls ignored what may have been a subtle hint and agreed to have Newmont participate. Therefore, in December, 1928, the new company, called Rhodesian Anglo American Limited, was formed, with Newmont Mining Corporation, in for 25 per cent, as the only American interest.

[1] Theodore Gregory, *Ernest Oppenheimer and the Economic Development of Southern Africa* (Cape Town, South Africa, Oxford University Press, 1962).

Oppenheimer became chairman of the new company, with Leslie Pollak as managing director in South Africa. Fred Searls was elected to the board. Rhodesian Anglo American took over the Rhodesian assets of Anglo American, including consulting engineering and management. Shortly thereafter, this participation in Rhodesian Anglo American (Rho Anglo, for short) put a severe strain on the Newmont resources at a very difficult time, but at the moment, Fred was fascinated by the seemingly unbounded opportunity for expansion in Africa.

Searls also turned his attention to South America, where Anaconda, Kennecott, and Cerro de Pasco were recording spectacular successes in Chile and Peru. Along with buying a large block of Cerro stock (which he sold a year or so later), Searls persuaded Judge Ayer to agree to a joint venture in exploration with United Verde Copper Company, of Jerome, Arizona. Robert Tally, head of United Verde, was also a Newmont director and a great friend of Fred's. Henry Dewitt Smith was then a vice president of United Verde, and from this association evolved the subsequent employment of Smith by Newmont, about the only really fruitful result for Newmont of the entire venture.

Formed in September, 1926, the new company, called New Verde Mines Company, was intended to be a vehicle for exploration in South America. Having decided on Peru as a starting point, Searls sent his friend George Scarfe off to Lima with instructions to build a staff and look for good bets.

In the process of looking, Scarfe visited Cerro's Morococha mine, where he chatted with Karl Schwegler, the superintendent, and toured the mine with a young geologist named Robert P. Koenig. Becoming impressed by both men, Scarfe attempted successfully to get first, Koenig, and later, Schwegler, to join him in the New Verde Mines exploration activities.

Harold Kingsmill, who was then running Cerro de Pasco in Peru, did not take kindly to this loss and promptly issued a separation paper which stated that Koenig could never again be employed by Cerro. Kingsmill also sent around an order saying that under no circumstances was Koenig ever to be allowed to set foot on any Cerro property again. Many years later, after Koenig had been elected president of Cerro de Pasco Copper Corporation, he happened on these documents in some old and forgotten files and, quite naturally, had them framed to be hung on his office wall.

New Verde's experience in Peru was not a profitable one, as it turned out. Exploration in Peru, before the days of helicopters, was quite difficult, and in addition, among the hundreds of opportunities presented to the company, the number of spurious promotions greatly outnumbered the genuine. Out of all the strenuous effort of Messrs. Scarfe, Schwegler, and Koenig, Newmont acquired only two properties: one was a prospect called Santander, and the other was a partially developed mine called Volcan both of which Newmont disposed of later when Fred Searls lost interest in Peru. Santander was brought into production in 1957 by St. Joseph Lead Company and is still operating profitably. Volcan also found other owners and entered production during World War II.

Had it not been for the onset of the Depression, beginning with the stock market collapse in October, 1929, Searls might have held on longer in Peru. As it was, he and Ayer decided in 1932 that Newmont's dwindling funds could be better employed elsewhere. New Verde's activities were shifted to other countries, and the cost of exploration in Peru of about $1 million had to be written off by the two partners.

In 1927, about a year after he had energized New Verde Mines Company, Fred Searls heard of a copper property in South Africa that before long diverted much of his attention away from Peru. This was the O'okiep orebody in Namaqualand, the only significant asset of the bankrupt Cape Copper Company Limited. American Metal Company had an option on the O'okiep orebody, but was dubious about exercising it. In 1927, Fred Searls visited O'okiep at the invitation of Otto Sussman of American Metal. Liking what he saw, Fred worked out a deal with American Metal whereby in 1928 a new company, South African Copper Company, was formed to acquire the assets of Cape Copper Company, which it did in 1931 for £159,240. Newmont directed the exploration campaign at O'okiep and developed plans to put the property into production.

The most complete geologic report on the Namaqualand copper deposits was written by Fred Searls, himself, in January, 1929. In it, Fred expressed concern about the absence of signs of strong copper mineralization in most of the area's outcrops. He wrote: "It makes me feel somewhat disappointed with the region as a whole, particularly as an important future copper field." Luckily for Newmont, his doubts weren't strong enough to stand in the way of the O'okiep project, which has had a successful history of very profitable production, with expanding ore reserves, for over thirty years.

In reading the several reports on O'okiep written for South African Copper Company, one meets many familiar names. For example, Ross Leisk, later manager at the Sunshine mine in Idaho; Phil Wilson, later with Lehman Brothers in New York; James Douglas, son of "Rawhide Jimmy" Douglas, who remembers well how tough a taskmaster Henry Krumb was in supervising the O'okiep exploration from New York. Also, there were Arthur D. Storke, who went to O'okiep in 1929, and Henry DeWitt Smith, who joined Newmont in 1930, and went to O'okiep in 1931 as one of his first assignments.

In transmitting his report on the estimated cost of putting O'okiep into production, Henry Smith told Fred Searls, "If you see a sale to Granby, Howe Sound, or Central in the immediate future at a price of $2 million or better, I am all for it. I am keen to have somebody else spend his money and time on South African Copper Company, perhaps because it is hard for me to visualize copper at over 13¢ per pound." Certainly, regarding O'okiep, Smith's crystal ball was as clouded as Fred's, because the property Smith would have sold for $2 million has since paid total dividends of about 125 times that amount. Furthermore, many years later, at the time of the Korean War, Henry Smith gleefully sold O'okiep copper at prices considerably higher than those he found so hard to visualize in 1931.

Probably the most useful report on the production possibilities of O'okiep was prepared by George A. Kervin, who went down in 1930 as manager. Kervin had been associated with Colonel Thompson since the Mason Valley days during World War I, where he worked with McNab in designing and building the Mason Valley smelter in Nevada. At O'okiep, Kervin did an excellent job on capital and operating cost estimates, although as it turned out, his studies served only as an indication to help guide future work.

Unfortunately, by the time Newmont and American Metal were ready to proceed at O'okiep, the world's economy had gone sour. By 1932, the price of copper on the London Metal Exchange had fallen to the equivalent of $0.09 per pound, and the plans for O'okiep had to be shelved. For five years they stayed there, gathering dust, until a firmer copper price in 1937 aroused interest once more in O'okiep's copper. What happened thereafter is told in Chapter 5.

Having thus, as he thought, started things rolling in Rhodesia, South Africa, and Peru, Fred Searls turned his attention back to the area that had never really left his mind or his heart. Having grown up in Nevada City, California, surrounded by the legend and the actual fact of one of the richest gold mining areas of the country, Searls' affection for, and faith in, the Grass Valley and the Mother Lode regions was one of the strongest influences in his life. Certainly he was thoroughly familiar with the operation of the Empire mine, one of the most consistent gold producers near Nevada City ever since its discovery in 1851. One can imagine the young Searls listening to the stories told by the miners who worked at the Empire, and resolving that some day, if ever he had the chance, he would acquire that mine, possibly he would even grow rich from it.

With Newmont's resources back of him, and with a mandate from Thompson to find new properties, Searls decided the time for action in California might be at hand. Accordingly, in 1927, he assigned Billy Simkins to examine properties near Grass Valley. Foremost among these, of course, were the Empire and the North Star mines. The Empire had produced something over $50 million in gold and gave indication of being far from worked out. Although the grade of ore at the Empire was said to have become somewhat lean, Simkins's examination indicated enough additional ore on the property to solidify Searls's interest in the mine. A disturbing factor, however, was that the two main veins of the Empire were found to be dipping toward property owned by the neighboring North Star mine. Searls filed this information and bided his time, awaiting an opportunity to acquire both properties.

In the meantime, Colonel Thompson had Henry Krumb evaluate Simkins's reports. In a short note scribbled on February 2, 1928. Henry Krumb told Thompson: "Do not believe Newmont should bother with this proposition. Even if we were successful in finding more ore and also successful in reducing costs, the return would not be really great, and while fussing with it, we would be sure to overlook something better. I believe your time would be spent more profitably in studying Newmont's investments with the idea of making switches." Note here that even Henry Krumb was beginning to show some evidence of speculative fever, although later he was one of the very few who sold out his holdings at the peak of the market in 1929.

H. S. Munroe, a mining engineer who had just joined Newmont in 1928, recorded this opinion on the Empire: "Hate to be a crepe hanger, but there's little here to make me feel that the economic result will be essentially different during the next ten years than during the last like period. Don't you think we can do better than that somewhere else?" It should be noted that these negative opinions were based on gold at $20.67 per ounce, and on this basis, the Empire and the North Star really did not look too attractive.

However, Fred Searls was nothing if not confident in both his own ability as a geologist and the possibilities of California gold mining. He remained determined to acquire as much of the area as he could, but was unsuccessful until an opportunity arose in the middle of 1929.

How Fred Searls brought the Empire and the North Star mines together is told in a long handwritten letter he sent to Colonel Thompson dated May 1, 1929. He wrote the letter on the Twentieth Century Limited somewhere in Ohio, when he and his brother Carroll were headed back to San Francisco because of the sudden death of their father, of pneumonia, on April 30.

To summarize the letter, Fred mentioned first that Newmont had been offered an option on the North Star mine for $500,000. After consultation with Simkins, Fred turned down the offer. It was well that he did, because soon afterward, Mr. W. B. Bourn, of the Spring Valley Water Company in San Francisco, suddenly realized that he had no further interest in the Empire mine, which he and his family had owned for thirty years. Bourn's only daughter had died rather suddenly, his wife had become ill, and he, himself, was already an invalid. Having known the Searls family all his life, Bourn thought of Fred as a possible buyer of the Empire and wired him an offer, subject to immediate acceptance, of the Empire mine outright for $250,000 cash. Fred saw this as the golden opportunity it was, and jumped at it.

Somehow or other, William Wallace Mein in San Francisco heard of the offer and immediately came to Bourn with the proffer of $300,000 cash. However, Bourn was a man of his word and turned down Mein's offer to stick by his proposition to Searls. Immediately thereafter, Searls incorporated the property as the Newmont-Empire Mines Company, a Delaware corporation, into which Newmont put $150,000 working capital.

Shortly after the announcement appeared in the San Francisco papers, the owners of the North Star Company approached Searls with the suggestion that they put their mine, plus $150,-000 working capital, into a new company, in return for which they would receive 49 per cent of the new shares. The main reason for accepting this proposal was that it would solve a possible legal problem regarding ownership of the two veins that appeared likely to pass from the Empire to the North Star.

Convinced that he had a bargain in hand, Fred accepted, and a second new company was formed; the Empire Star Mines Company Limited, of which Newmont-Empire owned 51 per cent. The other 49 per cent was divided among about 200 North Star shareholders. Newmont-Empire Mines Company was retained in being because Searls thought it might be useful somewhere else.

Formation of Empire Star Mines Company Limited certainly does not rank with the largest of the Newmont ventures, but it does rank high in its contribution to the company's stability. For one thing, the California properties were acquired at what turned out to be a bargain price. For another, the dividends paid to Newmont by Empire Star, though a modest sum of $3,216,210 from 1931 to 1943, sustained Newmont through the lean years of the Depression. For another, as Searls, himself, said in his letter of May 1, 1929, to Thompson: "I have come to believe that Newmont should do some operating. We need a place in this country to and from which we can ship foremen and superintendents. The Grass Valley mines are easy to run, and we now have the business organized, will not take my time and attention, nor that of others in New York after the first few months."

In keeping with Searls's expressed intention, Empire Star and its associated mines served as a training ground for a number of men who later assumed importance, either in Newmont's operations or in the operations of other mining companies. Among those Empire Star graduates who stayed with Newmont were Fred Scheck, Frank McQuiston, Ken Tatman, Jack Mann, Earle Currie, Bob Fulton, John Keenan, and Gene Tucker. George Kervin, who once managed the Nevada City mines of Empire Star and later O'okiep, was killed in an accident at Gray Eagle in December, 1941. Bob Hendricks, who was also a manager for Empire Star at Nevada City in California and with Northern Empire in Canada, eventually became manager in Cyprus for the Cyprus Mines Corporation. Dick Mollison, early in his career, was an engineer with Empire Star and later with

Idarado Mines in Colorado, but is now much better known as the man in charge of the exploration campaign that discovered the Kidd Creek base metal orebody of Texas Gulf, Inc., near Timmins, Ontario, and who now directs the subsidiary company that operates this mine.

Empire Star completes the list of Newmont's mining ventures in the 1920's, and it is now necessary to turn from Newmont, the miner, to Newmont, the speculator.

As a preliminary to a look at the speculative side of Newmont's activities in the twenties, it is helpful to review briefly Colonel Thompson's purpose in setting up the company. He wanted Newmont Mining Corporation to be a mining company in fact as well as in name. However, he did not think in terms of one or two mining enterprises. He had the rather unusual idea of having Newmont invest in several mining enterprises, at least some of which the company would manage as well as share in ownership. These active participations were to be financed by income, or capital appreciation, from a portfolio of mining and petroleum stocks, the shares of which would be readily salable when needed. This portfolio was to be restricted in nature because mining and petroleum stocks were presumably the ones that the staff of Newmont would be best acquainted with through their professional competence.

Thompson had selected his men intelligently, and the new company got off to an excellent start. Newmont's first dividend of $0.60 per share on the 430,000 shares outstanding was declared payable on October 15, 1925. Quoted on the New York Curb Exchange, the price of Newmont ran rapidly up and within two years had climbed to over $100 per share.

The mining and petroleum shares that Colonel Thompson had turned over to the company also soared in price. For example, in 1927, Newmont earned a fantastic $32.00 per share on the 452,760 shares then outstanding, largely because of having sold nearly all its Texas Gulf Sulphur shares at a very large profit.

Back in 1925, Newmont borrowed more than half a million

dollars from the Chase Bank and $1 million from J. P. Morgan for the purpose of buying stock in Kennecott and the Pacific Oil Company. By 1927, borrowing was no longer necessary. Much of the profit on the Texas Gulf sale went into purchase of more Kennecott stock. By 1929, Newmont's holdings of Kennecott had climbed to over 400,000 shares, so that the $5.00 dividend Kennecott was paying in itself about covered the $4.00 dividend that, at that time, Newmont was paying on its 503,224 shares outstanding.

It is easy to understand why the new administration of Newmont began to take greater interest in Wall Street than in the West. When one could buy a block of Noranda stock in 1924 for $100,000 and find that within two years it had become worth $1,550,000, as Newmont did, how could anyone get excited about a mining enterprise that might take several years of hard work to develop and then never pay off as well as this simple market transaction?

Although Colonel Thompson spent little time in the office during these years owing to his deteriorating health, he was not so ill that he could not think clearly about the future. Never having been dismayed by setbacks, he was not about to be stampeded by what looked to him like a runaway market. He knew that the prices of stocks, including Newmont's, had risen beyond all proportion to actual value. Beginning in 1927, Thompson had Henry Dodge quietly begin selling his own stock, putting the proceeds into bonds or holding them in cash, an action in which the prudent Henry Krumb eventually joined. Thompson was able to do the same with much of Newmont's holdings. The result was that by early 1929, Newmont had accumulated cash and bonds of around $35 million.

At just about that time, Thompson became almost entirely incapacitated, and his influence in the company lessened sharply. Lacking Thompson's steadying influence, Newmont began in May, 1929, an accelerated program of stock purchases, still entirely in mining and oil shares. At the same time, the company began selling fixed income securities. The result was that by September, 1929, practically all the $35 million had been reinvested, and the company was getting short on cash. Also, because of Fred Searls's commitment to invest in Ernest Oppenheimer's new company, the Rhodesian Anglo American Corporation, there were substantial payments coming due for purchases of Rho Anglo stock. For example, in October, 1929, Newmont's cash was down to just over $1 million, yet in November a payment was due for Rho Anglo stock of $1.5 million. Therefore, the company had again to authorize borrowing up to $3 million from the Morgan Bank, and had to repeat such borrowings in the next three years.

Yet no one was particularly worried. Despite the disquieting break in the market in October, 1929, Newmont could still show substantial earnings for the year, even though it made no profit at all in the fourth quarter. Newmont actually earned for 1929 a surprising $23.35 per share on its 504,424 shares, and although its stock touched an all-time peak of $236 per share in August, 1929, this did not seem extraordinary inasmuch as it represented a price-earnings ratio of only 10 to 1.

In the newspapers in 1929 writers were still referring to Newmont as having "the best mining engineers in the world, who tell the company's management what stocks to buy or sell." In other words "Newmont knows." It became obvious later that Newmont knew no better than anybody else. Like nearly everyone in the financial community in the fall of 1929, the men within Newmont, except for Dodge and Krumb, simply could not grasp the full extent of what was happening. An occasional market upturn buoyed their hopes. In fact, in 1930 Newmont was still granting certain new employees options on Newmont stock at $150 per share, which, as it turned out, was a nice gesture but one of no practical value.

Newmont's earnings dropped to $4.39 per share in 1930, and went on down in 1931 to a loss of $0.38 per share. In 1932, the loss increased to $1.23 per share on the 531,646 shares outstanding, but these were the first and last loss years in the company's history.

The price of Newmont's stock slid in roller-coaster fashion from its 1929 peak of $236 to its all-time low in 1932 of $3.875. Nonferrous metals, the main source of Newmont's income, collapsed in price. Copper fell from a 1929 high of $0.178 a pound to a Depression low of $0.04775 a pound in January, 1933. Lead and zinc hit unbelievable bottoms of $0.0275 and $0.0253 a pound, respectively, in July and May of 1932. Silver dropped from $0.56 per ounce in 1929 to $0.26 in February, 1933. At such prices, nonferrous metal producers passed dividends right and left, and Newmont's dividend income almost disappeared, reaching a low of $177,450 in 1933. The company's stockholdings, which in 1929 had a market value of $52 million, were worth only $9 million by 1933. Struggling with debt, surrounded by investment wreckage, Newmont's managers faced their moment of truth, not only within the company, but in their personal fortunes as well.

It is a rare individual who can accept gracefully and immediately the end of a love affair or a disappearance of prosperity. It took everyone in Newmont until 1932 to accept the fact that a fundamental and long-lasting change had taken place in the country's economy, and that the days of quick profits in a bull market were over for a long, long time. The personal losses suffered by Searls, Kraft, and others, as well as the Newmont corporate losses, made such a deep and lasting impression that never again has Newmont been regarded by its managers as, in any sense, a speculative vehicle.

To sum up, then, in its first decade, Newmont looked at hundreds of prospects all over the world, but became involved in actual mine development only in Canada, Peru, South Africa, and California. Of these attempts at mine operation, only Empire Star in California resulted in cash income in Newmont's first ten years. Mason Valley was pre-Newmont. All of Newmont's total income of about $50 million through 1932 came from dividends, interest, fees, and capital gains on sale of securities. There was not one penny derived from any mining activity actually initiated by Newmont. The company's heavy dependence on the stock market and the mining operations of others had not served it well. Only in its next decade, and thereafter, did Newmont become the true mining company Colonel Thompson had always wanted it to be.

Empire Star miner, 1929

4   A New Cast and Recovery . . .

Convinced at last that the good times were gone, Newmont's men wasted no time in remorse or recrimination. Faced with an actual fight for survival, they set about cutting costs, they looked for men with whom they could build a stronger staff, and they began to reorient their exploration efforts toward gold.

Because Newmont's operating costs in 1932 consisted mostly of salaries and office expenses, a cost-cutting drive had to hit salaries first. Judge Ayer had already cut his salary; in fact, in 1931 he gave up his salary entirely and never thereafter had it resumed. All other salaries over $10,000 were reduced by 20 per cent, and all under $10,000 were cut by 10 per cent. This sounds worse than it actually was, in that the cost of living went down at the same time.

By this and other measures, Newmont cut its costs for 1932 by 16.7 per cent below those for 1931, which already represented a 30 per cent reduction below the level of 1930.

Costs could also be cut by reducing exploration and development. In 1932, Searls pulled New Verde out of Peru and, in agreement with Newmont's O'okiep partners, shut down the mine development program in South Africa. Yet a mining company can't live without exploration. Neither can a company short of funds take too many exploration gambles. To Searls, pleased as punch by the way Empire Star was working out, there was only one way to minimize exploration risks in the early 1930's, and that was to go after more gold properties in California and in Canada.

In 1931, Newmont bought the Murchie gold mine on the Mother Lode in California and the Original 16 to 1 mine near Alleghany, California, a small, very high-grade gold property Searls remembered from his early exploration days with Joe Thorne. As it turned out, it was neither original nor all that high grade, the best ore having been mined years before, but Searls did have a lot of fun with it. In 1932 and 1933, Newmont came across two Canadian properties, Island Mountain in British Columbia, and a prospect that became Northern Empire Mines Company in Ontario.

Even with gold at $20.67, these properties looked attractive, mainly because of gold's fixed price as compared with slumping base metal prices. But in February, 1934, Fortune smiled on Newmont in the unlikely person of Franklin D. Roosevelt. In that month, as one of his moves to counteract the effects of the world-wide depression, FDR got Congress to set the price of gold at $35.00 per ounce.

Practically speaking, the Depression, for Newmont, ended right there. Empire Star, which had been producing a little over $2 million in gold a year, immediately saw its output jump to a value of $3.5 million without mining a single additional ton of ore. In turn, Newmont's dividend income rose to $1,109,470 in 1934 from $177,450 in 1933. By additional borrowing and judicious sale and purchase of securities, Newmont was able to build up the size and market value of its portfolio from about $9 million in 1932 to nearly $30 million at the end of 1934.

Here was a perfect example of serendipity at work. The company (Empire Star) that Fred Searls had put together in 1929, against the advice of his associates and for an investment of less than $500,000, became the only strong support of Newmont during the Depression. Searls obviously did not foresee the Crash and the Depression, although Thompson, Dodge, and Krumb apparently did. But certainly none of them could have forecast gold at $35.00 by February, 1934.

Searls thought he was buying two rather long-shot gold mines in a part of the country he loved. What he actually bought was the future security of Newmont. It could be that no subsequent accomplishment ever made Fred Searls happier than this one. Obviously, he could not have planned what actually happened. Rather, he had acted largely out of faith in the mines and the area he knew so well from childhood. The strength of his faith was well matched by the richness of his reward.

Under the spur of the new high price for gold, Fred Searls and Newmont then went all out on a widespread program of gold mine exploration and development. This concentration on gold understandably dominated Newmont's policy planning through the 1930's. To handle this gold mine development program and the resulting management problems, Newmont acquired some new staff members during the early 1930's, all of whom profoundly affected the company in later years.

The first, and certainly a most important such addition to Newmont's staff was Franz Schneider. Joining the company in 1930, Schneider combined an engineering background with reportorial skill and keen financial judgment. Educated at Massachusetts Institute of Technology, Schneider, after several varieties of employment, became financial editor of the New York *Post*. After several years he moved to the New York *Sun*, the principal owner of which was Tom Lamont of J. P. Morgan & Co., the man who had been such a close friend of Colonel Thompson.

As the Depression deepened, Lamont became gravely concerned about Newmont, not so much for its ability to find and develop mining properties, but in its ability to weather the financial storms that Lamont saw ahead. Knowing of Franz Schneider's financial acumen, Lamont succeeded in having him brought into Newmont in 1930 as a vice president.

Schneider wasted no time. In a series of transactions, he was able to build up fairly soon a fund of about $3 million in cash that he intended to use as a defense against possible future financial difficulties. A complication arose, however. Years earlier, Colonel Thompson had become a shareholder in Continental Oil Company in a manner described in Chapter 15 on Newmont in the oil industry. For present purposes, it is enough to mention that when Fred Searls had resigned as Newmont's representative member on the Continental Oil board, he had nominated Franz Schneider as his replacement.

Shortly after joining the Continental board, Schneider went to Colorado and elsewhere in the West to inspect the Continental Oil holdings. He was gone several weeks, and on his return to New York, he found that his carefully accumulated kitty of $3 million had been reinvested in stocks that he could not find it in his heart to approve.

This was too much for Schneider. As a result of his urging, Tom Lamont and Albert Wiggin were able to persuade the Newmont board to create in October, 1931, a new organization under which management responsibility for the company would be handled by three men, Judge Ayer, Fred Searls, and Schneider.

For most companies, such a triumvirate with its consequent division of management responsibility might well have been disastrous. For Newmont, however, it seemed to have been exactly suited to the talents of each man. Judge Ayer continued as he had in the past to look over each activity of the company, bringing to bear the experience, the judgment, and the legal mind that had been so important to Newmont. Exploration became the sole responsibility of Fred Searls, at which he had already demonstrated superior ability. Franz Schneider was to handle sales and purchases of securities for the company, which he did thereafter most effectively. Referred to within the company as "the three-headed dog," the triumvirate managed Newmont's activities from 1931 until January, 1947, at which time Judge Ayer became chairman of the board, with Fred Searls replacing him as president and Franz Schneider continuing as executive vice president.

The second new member of Newmont's management who, like Franz Schneider, greatly influenced the company from almost the first day of his advent, was Henry DeWitt Smith. Born in Connecticut in 1888, Henry Smith graduated from Yale, receiving an E.M. degree from the Sheffield Scientific School in 1910. He then joined the exploration firm of J. E. Spurr as a geologist and did field work for Spurr in the western United States and in Mexico. No better training for a young geologist could have been found than working with Spurr, partly because of Spurr's widespread interests and partly because of Spurr's tremendous reputation in the mining industry.

In 1914, Smith decided to obtain experience in mining, and he took a job as shift boss with the Kennicott Mines Company in Alaska. In his three years at the property, Smith was promoted from shift boss to mine superintendent and to assistant manager. In 1917, he went to the United Verde Copper Company at Jerome, Arizona, as mine superintendent. In the seven years that Smith spent at United Verde, he increased his familiarity with copper orebodies, he was promoted to general superintendent, he became acquainted with Fred Searls, and he developed training programs for mining engineers that are still in use and are still regarded as models of their kind. In developing these programs, Smith also developed a liking for training and helping young engineers. Throughout his life, he retained a strong interest in assisting the careers of young mining men.

In 1924, Smith came to New York as a member of the industrial department of the New York Trust Company, and for three years he busied himself with investigation of various industries, none of which had anything to do with mining. Although he found this activity profitable, he did not enjoy being unable to use his experience in the mining industry, and in 1927, he returned to work for United Verde, but this time as vice president and sales manager in New York. Three years of this was enough, however, and in late 1929, he joined Fred Searls and Newmont as staff engineer and a member of the Newmont board. Smith became a vice president in 1947.

In time, Henry Smith came to represent Newmont in the minds of other mining industry members, almost as much as did Fred Searls. A tall, handsome, athletic-looking man, Smith's ruddy countenance and shock of closely cropped gray hair became a familiar sight at mines both in the United States and in foreign countries. No doubt he had his more serious moments, possibly even showing irritation at some vexing problem, but based on fairly frequent observation over a period of years, the recollection that comes immediately to mind is that of twinkling eyes and a friendly smile. He was willing to listen to other people's problems and had an unusually strong ability to think through to the heart of the matter, coming up with a reasonable solution. His helpful influence was felt not only in Newmont, but in other companies, such as St. Joe and Phelps Dodge, where he served as a director.

Also entering Newmont in the late twenties was William T. Smith, whose acquaintance with Colonel Thompson and Newmont goes back to the early days in Butte, where Smith's father was a neighbor and friend of the Thompson family. When Thompson helped organize Nevada Consolidated, he asked Smith's father to assume responsibility for the various community services at Ely, Nevada, where the mine was located. Naturally, William Smith became interested in Newmont, and in 1927 he joined the company in the treasurer's office. Smith eventually followed Henry Dodge and Gus Mrkvicka as treasurer of Newmont. Following his retirement in 1967, Smith has continued as one of the administrators of The Boyce Thompson Institute for Plant Research in Yonkers.

This history of Newmont's early years owes much of its material to Bill Smith and to Warren Publicover, who came to

Newmont from Nova Scotia in October, 1926, out of Dalhousie University at Halifax. The son of a sailing master out of Lunenburg, Nova Scotia, Publicover was a family friend of the Dodges and was encouraged by Dodge to try his luck in New York. At Newmont, Publicover worked for Ayer and Krumb for the most part, and also for McNab.

Publicover was responsible for supplying Newmont with a third Nova Scotian in the person of Earle K. Currie, an old friend who was, in 1927, working in Lunenburg for a branch of the Royal Bank of Canada. Unhappy with his compensation and his prospects, Currie was quite receptive to Publicover's suggestion that he join Newmont in New York, which he did on January 1, 1928.

Currie began work in Newmont's accounting department under Henry Dodge, and he remembers warmly Dodge's encouragement during his first difficult weeks in a strange office and an even stranger city. In 1931, Fred Searls had Currie go out to San Francisco to help transfer the Newmont office there from the Mills Building up to Grass Valley, which became the headquarters of Empire Star. Currie thoroughly enjoyed his stay at Grass Valley, as well he might, for it is one of the most attractive surroundings in the world for a mining camp.

Returning to New York in 1935, Currie resumed his work in the accounting office under Henry Dodge, where he continued as assistant to the treasurer until his retirement in 1966. Here there developed a kind of divided responsibility, with the Newmont accounts being handled by William T. Smith and Earle Currie, and the Magma and various Newmont subsidiary accounts being handled eventually by Gus Mrkvicka and Walter P. Schmid. Although both groups reported to Henry Dodge, neither knew what the other was doing, and any interchange of information was discouraged. This independence of operations between Newmont and Magma grew more pronounced over the years and reached a kind of climax in 1948, when the boards of the two companies were entirely separated in the course of the effort to finance the San Manuel project. (See Chapter 9.)

Among the members of the Newmont staff most familiar

with the early days in California was Fred Scheck. Fred was born in Germany and grew up with a desire to become a farmer. He was interested enough in modern farming to desire to learn the most modern techniques of animal husbandry. He had heard that the Iowa State College at Ames was the best place to learn such techniques, and Fred therefore came to the United States and enrolled in the college, later moving on to the California Agricultural College at Davis. Fred soon discovered that he didn't care for either the college or for farming as it was then done in California.

Changing his objective, Fred enrolled in a business college in Sacramento, and after graduation he took a job at the Murchie mine on the Mother Lode in 1929. At that time, American Foundation Company owned the Murchie, but in 1931, the property was sold to Empire Star. Fred Scheck stayed with Newmont until his retirement in December, 1970. He spent several years in Africa with O'okiep and, in later years, returned to the New York office where he was active in the accounting affairs of the organization and in the financial analysis of new ventures.

Strong influence in the mining activities of Newmont appeared in 1934 in the person of Marcus D. Banghart. Born in Nebraska. Banghart graduated in geology from the University of Nebraska. For several years after graduation, he worked for Anaconda in Butte, Montana. He then went to Ecuador for South American Development Company, a gold producer, becoming mine superintendent. He left Ecuador at about the time the Depression was beginning, and thereafter he was employed at five different mining properties, all of which shut down successively.

Just prior to his association with Newmont, Banghart worked with the Lucky Tiger Mining Company, at first doing mine examinations out of their Oakland, California, office. His stipend at the time was $5.00 per day. Later, for the same company, Banghart went to a mine at Mokelumne Hill. Here Banghart performed all the functions of mine and mill management, except accounting, for $225 per month. The company wanted him to do the accounting as well, but Banghart refused.

In 1934, Henry Smith, with whom Banghart had become

acquainted previously, offered him the job of manager of the Island Mountain mine in British Columbia at a salary of $450 per
month. In the Depression, this was an extremely attractive offer
and Banghart took it.

In 1939, Banghart was transferred to the Berens River mine
of Newmont in Canada, where he stayed until 1941, when
Henry Smith again approached him, this time to ask if he would
be interested in going to Africa, where the O'okiep property had
just entered production. Banghart assumed charge of O'okiep,
and later Tsumeb as well, returning to New York in 1954 as a
Newmont vice president, with responsibility for all the mining
properties for which Newmont was the manager.

A former football player, Banghart has the rugged appearance and brusque manner that go with the popular image of the
game. Hidden behind this rough surface, however, is a keen and
broad intelligence coupled with a sense of fair play and a basic
kindness that, when it is discovered, turns a wary stranger into a
lifelong friend. Broad gauge in judgment and simple in approach, Banghart, like so many Newmont men, sensed the nub
of a problem without the necessity of elaborate analysis. Banghart's approach to management may have been old-fashioned,
but it was tremendously successful, the best evidence of which
lies in the devotion and loyalty of all the men who worked with
him. Banghart, himself, has been completely loyal to the company, and his high integrity was the underlying cause of the respect all men felt for him. Retiring in 1968, Banghart now lives
in South Africa, where he is still retained by Newmont as a consultant.

Still another member of the new cast of characters of Newmont was Wesley P. Goss. Born in Kansas, Goss studied mining
engineering at the University of California at Berkeley, graduating in 1922. Following summer experience as a miner in various
Mother Lode properties, Goss went to work with United Verde
Copper Company at Jerome, Arizona, where he filled successively positions as engineer, shift boss, foreman, and assistant superintendent. In 1934, he transferred as general superintendent

to the Park City Consolidated Mining Company in Utah. In 1935 he joined Newmont's exploration staff and worked under Phil Kraft in a search for lead ores jointly financed by The Ethyl Corporation and Newmont.

In 1937, Goss was offered the position of mine superintendent at O'okiep, when development of the mine was begun. Goss stayed at O'okiep as mine superintendent, working four and one-half years without a vacation. He returned to the United States in 1941 and went to California for Newmont as assistant general manager of the Gray Eagle Copper Company in northern California. After the Gray Eagle orebody was worked out, Goss was sent to make an examination and a report on the Magma mine at Superior. He did this so capably that in July, 1944, he was made general manager of Magma Copper Company. Goss became president of Magma in December, 1953, serving in that capacity until January, 1972, when he became chairman of the board. Magma and Newmont owe much to Goss's ability both as a mining engineer and as an administrator. The San Manuel development, which Goss supervised, is one of the largest scale projects with which Newmont has been affiliated.

An energetic, forceful man, Goss is not one to parade his innermost thoughts, but he leaves no doubt as to his opinions on any subject under discussion. When his mind is made up, he can cut off an argument with a curt, flat statement. On the other hand, his friends know him to be a kindly, generous man deeply concerned with the problems of the people of his community and his state. Not even all his associates know that Wes Goss has served capably for years on the Board of Regents of the two Arizona universities, most recently as board chairman. This work he has carried on in what is probably the most difficult period in the history of these institutions. Impatient with slackness in performance and strongly loyal to his company and to his profession, Wes Goss seems to have been admirably suited to the demanding job in which circumstances and Newmont have placed him.

These are some of the men who, in Newmont's second decade, faced up to the problems of recovery from the Depression. Base metal mining was flat on its back. The petroleum industry was not much better. The only metal whose price remained stable was gold. An obvious move for Newmont, therefore, was to concentrate on gold mining, and the obvious channel through which to seek new properties was Empire Star.

The first such acquisition by Empire Star had been the Murchie mine in 1931. Under its former owner, the Murchie produced about 150 tons per day of sulfide gold ore, but after Empire Star bought it, production was doubled. Manager of the Murchie mine under Empire Star was George Kervin, just back from the O'okiep exploration in South Africa. Following the increase in the price of gold, Empire Star went after other nearby properties, eventually putting together a group that included the Zeibright, and the Pennsylvania and the Browns Valley mines.

Thus there were six operating mines in the Empire Star group. Fred Nobs, an old friend of Fred Searls who had come in with the original deal by which Empire Star was formed, became manager of the Empire, North Star, and Pennsylvania group with headquarters at Grass Valley. George Kervin was made manager of the other division, including the Murchie and the Zeibright mines with headquarters at Nevada City. When Kervin returned to O'okiep in 1937, Bob Hendricks, who had been chief engineer of the Murchie and manager of Northern Empire, became manager in Nevada City. Fred Nobs had Jack Mann as assistant manager in Grass Valley, and when Nobs retired in 1939, Mann became manager at Grass Valley. In 1942, when Hendricks went up to Gray Eagle, Mann became general manager of Empire Star, a position he held for twenty years through the remaining life of the company.

By 1934, the Empire Star group of six mines employed something over 1,000 men. One of these men was Frank W. McQuiston, Jr., a young flotation metallurgist who graduated in 1931 from the University of California's Hearst School of Mines at Berkeley. Associated with practically all of Newmont's metallurgical developments for the past forty years, McQuiston began work in 1931 for the Fred Bradley interests at the Spanish mine north of Nevada City. For a short time that year, the Bradleys loaned McQuiston to Empire Star to help develop a means of floating gold-bearing pyrite in a new type of flotation cell. When the Spanish mine closed, McQuiston worked for a while at the Argonaut mine, then went broke trying his hand at leasing near West Point, California. Finally, he came to work as an assayer and mill hand at the Empire mine, where he had an experience that brought home to him what the Depression really meant.

After he had worked in the Empire cyanide plant for a reasonable time, McQuiston braced Fred Nobs for a raise. Nobs pulled out a file folder about two inches thick and said, "Frank, here are over a hundred applications from engineers all over the world. They would all consider it a privilege to live in Grass Valley and work for Empire Star."

"Oh," replied McQuiston. "Thank you, Mr. Nobs. I hope you won't dock me for the time I spent coming to see you." From this beginning at a wage of $5.00 per day, McQuiston progressed to the point where, a few years later, he supervised development of a flow sheet for Empire that replaced the former practice of stamp mill crushing to 28 mesh, followed by plate amalgamation and Wilfley tables. The old system was expensive and open to easy theft. The new flow sheet included stamp mill crushing to only ⅜ inch, followed by jigs, continuous amalgamation, and flotation. Theft no longer was a problem, costs were reduced, and gold recovery was increased.

As a sidelight, McQuiston remembers that while this new flow sheet was being developed, Plato Malozemoff, who had suggested the jigs, called several times at the Empire and North Star mills in an effort to convince McQuiston that he should use a new type of flotation machine just developed by Pan American Engineering Company, Malozemoff's employer at the time. The future president and chairman of Newmont failed to impress McQuiston, who stuck with his old machine, although continuing to use the Pan American jigs.

The Zeibright mine, which was producing 1,000 tons of

ore a day running about 0.09 ounce of gold a ton, had to shut
down in 1939 when its tailings dam slid into the Bear River and
clogged the intake of the Pacific Gas and Electric hydroelectric
plant downstream. The Murchie mine ran out of ore at about
that time, but the North Star, the Empire, the Pennsylvania
and the Browns Valley mines produced all during the 1930's and
were only closed down in 1942 by the gold mine closing order, L-208, when the United States entered World War II. The
Empire and the North Star mines were opened again after the
war, but by then rising costs had caught up with the fixed price
of gold, and these famous old properties were eventually closed
and the Empire Star Company was merged into Newmont in
May, 1959.

Another Newmont venture in gold mining in the thirties
involved the Getchell mine in Nevada. Noble Getchell, a Nevada banker, had grubstaked two prospectors early in the 1930's,
and by 1935, the two had come up with a promising prospect.
To finance it, Getchell went to his former associate, George
Wingfield. At the time, Wingfield was unable to help on his
own, inasmuch as twelve of his chain of banks in Nevada had
closed early in the Depression and he had been forced into
bankruptcy. However, Wingfield, being a close friend of Bernard Baruch since their association at Goldfield, Nevada,
reached Baruch by telephone and asked for a $1 million loan.
Baruch was willing to send a check at once, but Wingfield insisted that he should have a competent engineer look at the
property first.

Baruch sent for Fred Searls, who was in Africa at the time
and unable to come immediately. By the time Fred Searls got to
Nevada, an adit had intersected high-grade ore, and the property
looked so good to Searls that not only did he recommend it to
Baruch, but he brought Newmont in as well.

At first, the Getchell was extremely successful and profitable. The company built a 1,000-ton-per-day cyanide plant to
treat oxide ore averaging $7.00 per ton in gold. Extraction was
over 90 per cent and operating costs ran about $1.00 per ton of

ore. The Getchell paid handsome dividends and everyone profited.

Once the oxide ore was gone, however, the sulfide ore proved quite refractory, and attempts to achieve success in cyanidation through a preliminary roast were never very successful. Frank McQuiston served as Newmont's representative and consultant to the Getchell Company from 1939 to 1947, when Newmont sold out. Fred Wise, the Getchell mine superintendent, joined Newmont as general manager of the Idarado property in Colorado shortly after World War II.

As profitable as the Getchell was for a period, the Empire Star group stands out clearly as the mainstay of Newmont after the Crash. From 1929, when Fred Searls established the company until it finally closed down in 1959, Empire Star produced approximately $40 million in gold. Total production of the Empire and North Star mines from 1850 until closure was in excess of $130 million. Aside from the fact that it was Newmont's main support during the Depression, Empire Star is of interest because it was organized and operated in typical Newmont fashion. Although by 1934 it was not a majority shareholder in Empire Star, Newmont did arrange to have the company begin acquiring and maintaining a portfolio of securities in other companies, including railroads and utilities, as well as mines. These securities were chosen to be readily marketable and capable of producing a reasonable income, pending the time when they might be sold to provide funds to finance a new mining property for Empire Star. This, of course, is a basic Newmont philosophy and, in essence, the manner in which the company still operates.

Thus having come through the pit of the Depression somewhat chastened but with renewed confidence in themselves and with a firm belief in the validity of their approach to mining, Newmont's executives again sought renewed growth in the reawakening mining industry.

Handling blister copper, O'okiep

5 . . . Reopen the Road to Growth

With the new price of gold as an incentive, Newmont's first worthwhile acquisition in 1934, outside California, was a gold property in Western Ontario that became known as Northern Empire Mines Company Limited. Newmont spent about $750,-000 in purchasing the property and equipping it. Beginning production in 1934, Northern Empire, by 1941, had produced 142,000 ounces of gold and had paid dividends totaling $1,-292,000 in Canadian funds.

As manager of Northern Empire, Fred Searls sent up Bob Hendricks, who had been chief engineer at the Murchie. Shortly after he took over the property, Hendricks experienced one of Fred Searls's characteristic attempts at economic operation. A Searls foible was his positive passion for equipping a mine property with used machinery, hoping thereby that the lower capital cost would outweigh any possible increase in operating and maintenance costs. In addition, Fred believed firmly that inasmuch as stamp mills had been used successfully by Empire Star, all gold mills should use stamps. The result of all this was that Hendricks received word there had been consigned to him a complete stamp mill that the Newmont engineers had unearthed from an abandoned property somewhere on the Mother Lode in California. The customs people held up the shipment at the border for several weeks because they had never heard of a stamp mill and couldn't find the machinery listed in any of their catalogs. Never having thought much of stamps, Hendricks was in no hurry to get the shipment released, although he did eventually use the equipment.

The Northern Empire mine had at one time been considered too low grade to be attractive, but Newmont was fortunate in discovering, shortly after taking it over in 1934, a vein 2 feet wide carrying 2/10 ounce of gold. This vein was vertical, but was cut off by a flat diabase sill 600 feet thick. Most of the ore from the Northern Empire came from above the sill. Years later, exploration disclosed some ore below the sill, but a diminishing grade forced the property to close in 1941.

Also in 1934, Newmont took over a promising gold property, the Island Mountain mine in British Columbia, that Fred Searls had first considered in 1932. To decipher the complicated geological structure at Island Mountain, Fred Searls assigned a brilliant young geologist named P. C. Benedict, who thereafter remained with Newmont for the rest of his working life. "Benny," as he was affectionately called, had an extraordinary ability, not only in geology but also in endearing himself to everyone he met. There is hardly a mining area in the world that Benedict has not visited, nor any such mining area where he is not fondly remembered. His activities everywhere were noteworthy, but were probably most outstanding in Canada, Arizona, Australia, and Africa. Benedict worked for Newmont until he retired in 1967. He died two years later.

At Island Mountain, the gold was found in a narrow band of limestone in an overturned fold. Zones of pyrite replacement in the limestone carried 1 to 3 ounces of gold. In the footwall of the limestone fold there were veins normal to the structure carrying about 3/10 ounce of gold. When Newmont took it over, Island Mountain had a mill with a capacity of about 50 tons a day, recovering the gold by cyanidation. Henry DeWitt Smith had been given the Island Mountain as one of his responsibilities, and it was Smith who invited Marcus D. Banghart to be manager of Island Mountain. Banghart remained there until 1939, by which time he had increased production to about 150 tons per day with an annual profit of about $250,000 on the operation.

The third Canadian gold property operated by Newmont was the Berens River Mines Limited near Favorable Lake in Manitoba. The property was brought to Newmont's attention in 1935 by John Drybrough, who years later became president and eventually chairman of Newmont Mining Corporation of Canada Limited, a wholly owned subsidiary of Newmont. Born in Edinburgh, Scotland, Drybrough came to Canada in 1908. After graduating from high school, he took a job with the Mond Nickel Company. After a short period of employment, he left to join the Canadian Army. He was in the army until 1919, when he returned to Canada and studied at the University of Toronto, graduating as a mining engineer. Drybrough then returned to the Mond Nickel Company, but he left the company after several years to set up a consulting office in Winnipeg from which he did exploration work as a consultant, in the course of which he came across Berens River in 1935.

Having met Fred Searls a short time previously, Drybrough thought he might be interested in Berens River. Searls was quite impressed by what he heard and sent Benedict from Island Mountain up to Manitoba to examine the prospect.

With a kind of enthusiastic, all-or-nothing approach, Benedict placed a line of explosives in the muskeg over what he hoped was the hidden outcrop of a vein. The resulting explosion blew muskeg in all directions, revealing an almost flat exposure of bedrock containing a vein 10 feet wide that proved to have an average grade of 12 ounces of silver and 3 ounces of gold per ton. Certainly few veins in history have been exposed so rapidly, so completely, or with such gratifying results.

Berens River was brought into production in 1936, with Henry Smith, as president of the company managing it, and with Banghart becoming general manager in 1939. Change of scene, however, is part of every mining man's experience, and in 1941, Banghart went off to O'okiep in South Africa, and in 1942 Henry Smith went down to Washington to work with the Metals Reserve Company. Thereupon, John Drybrough became president of Berens River and directed its activities for the next nine years. By May, 1948, Berens River had worked out practically all its ore reserves, and it operated thereafter as an exploration company covering various prospects in Canada. In 1959, Newmont merged Berens River into Newmont Mining Corporation of Canada, after buying out the public shareholders in Berens River on quite generous terms. At about the same time, Island Mountain was sold, and the proceeds from its sale were also placed with Newmont of Canada.

About four miles from the Northern Empire Mine, Benedict, in 1943, examined a prospect that eventually was operated under the name of Undersill Gold Mining Company. This name arose from the fact that the ore on the property was found under the same diabase sill that caused trouble for Northern Empire. The Undersill orebody was opened by a shaft that was sunk under tremendous difficulty caused by very heavy flows of hot, salty water. The water flow disappeared once the shaft had cut through the sill, and at the 1,800-foot level, a drift driven over to the vein discovered it to be about 3 feet wide, running an ounce of gold to the ton. Unfortunately, the vein had a very short strike length. Opened in 1944, jointly with a neighboring company, Undersill produced only until 1949.

Other Canadian gold properties in which Newmont was interested in the late 1930's or early 1940's were the Magnet Consolidated Mine, which was characterized by narrow veins and small production; the Tombill gold mine which was operated between 1935 and 1946; and the Sachigo River Exploration

Company, which was operated by a syndicate with Newmont as one of the members. Located in Northern Ontario, the Sachigo River property is said to have produced about $600,000 in gold between 1935 and 1941. Incidentally, it was at Sachigo River that Newmont first encountered Eldon L. Brown, who was manager there. Brown and Newmont met again, more productively, with the Sherritt Gordon development in 1952.

All in all, Newmont's ventures into gold mining during the 1930's provided a profitable antidote to the setbacks of the Depression. However, not all Newmont's experiences during that decade were so successful. In 1936 an exploration party directed by Fred Searls came across a copper property near Jarbidge, Nevada, called the Gray Rock, that had at one time been quite a large producer, largely because of the high grade of the ore found in the Long Hike vein in the Gray Rock mine. Having run into a fault that cut off the vein, the owners had finally been forced to shut down the mine because they could not relocate the vein beyond the fault. In 1937, Newmont bought the property and began development work on the strength of Fred Searls's opinion that the lost vein could be found.

The initial work at the Gray Rock consisted of deepening an existing shaft for exploration at depth. All went well until the shaft was down about 1,000 feet, when warm water began to flood the workings. Before long, the flow had increased to about 5,000 gallons a minute. Additional pumping and power-generating equipment had to be supplied. In the meantime, diamond drilling had located the lost Long Hike vein just where Searls had thought it would be. Its grade of copper was just as high as the ore originally mined in the Gray Rock.

Mining can be among the most tantalizing and frustrating activities in the world. With the certain knowledge that the prize was there for the taking, Newmont's engineers redoubled their efforts to keep up with the influx of water. How ironic that this battle against water took place in one of the driest regions on the American continent.

News of Newmont's fight against the flood spread well beyond the neighborhood of Jarbidge. At one point, Fred Searls received a letter from an elderly mining engineer who said, "I used to get legal counsel from your dad back in Nevada City in 1910." He went on to say that he had been "pulling for you ever so hard to make good at Jarbidge for three reasons: first, I want to see you and your company make good. Second, I always contended the Gray Rock mine was not done for; third, I know there's plenty of ore still there."

Searls was deeply moved by the old engineer's comments. He wrote back: "Sir, I remember working with Winchell and John Gray on the Big Jim case. Poor old Judge Lindsay got so mad at Winchell in court that his blood pressure mounted to where he had an internal hemorrhage that killed him. I would like to meet you again and buy you a drink."

In the end, all the money, good wishes, and engineering skill that Newmont could accumulate just were not enough to handle the pumping problem at Jarbidge. By the time Newmont gave up, after three years of trying, the engineers estimated that they had pumped something like 7 billion gallons of hot water up out of the Gray Rock shaft. In 1941, the mine was closed and the Gray Rock company was, appropriately, liquidated.

This seems a suitable point to mention that Newmont, at other properties, has pumped a great many more gallons than those that flowed out of the Gray Rock.

In 1953 at Tombstone, in Arizona, an attempt at exploring a gold property met the same experience. One of the geologists on the job expressed his frustration by recalling the Tombstone project as having developed "a mighty good water mine." The citizens of Tombstone could have used the water if they could have afforded the pumping job. Newmont retained a stock interest in the property for about ten years in the hope that someone, Tucson for example, might want the water, but no one did.

Newmont met a third such experience at Eureka, Nevada, beginning in 1963. This was the Ruby Hill property, a high-grade lead-zinc-silver orebody that Fred Searls had carried in the back of his mind ever since he examined it as a geologist for U.S. Smelting back in 1924, before his connection with Newmont. Ruby Hill was owned by Richmond-Eureka Mining Company, in which U.S. Smelting had a 70 per cent interest. The original discovery at Ruby Hill was a rather shallow orebody cut off by a fault. On examining it, Searls came to believe that there was at depth a downthrown segment of the orebody that would be worth looking for. Thayer Lindsley, of Ventures, Limited, a Canadian company, also thought well of Ruby Hill and did some drilling under a lease from Richmond-Eureka taken by his subsidiary, Eureka Corporation, in 1938. Results were encouraging, and during the Second World War, the U.S. Bureau of Mines did enough drilling on the property to disclose a deep orebody where Searls and the others thought that it would be. Thayer Lindsley then proceeded to active development. In 1941, the Eureka Corporation started sinking a new shaft, called the Fad Shaft. It bottomed at 2,415 feet in late 1947, and crosscuts were driven toward the orebody, on the 2,250-foot level. Some 50 to 70 feet from the shaft, one crosscut hit the main fault, which suddenly and catastrophically let go its gouge and flooded the shaft. The drift crew barely escaped with their lives. After several attempts at unwatering the shaft, Eureka Corporation found no way to handle the water flow, and the Fad Shaft lay dormant from 1953 until 1963.

In the latter year, Newmont, in partnership with Cyprus and Hecla, formed Ruby Hill Mining Company, which leased the property from the Richmond-Eureka Mining Company, and tried again to reach the Ruby Hill orebody through the Fad Shaft—so tantalizing was the conviction of Fred Searls that the high-grade ore could be extracted profitably. Searls was joined once more by Bob Hendricks, recently retired from Cyprus, and Bill Love, of Hecla Mining Company, in the opinion that recovery of this flooded mine was possible. They based their hopes on recent developments in technology that had successfully overcome heavy water flows by means of deep directional drilling by a Texas oil-drilling contractor, combined with use of chemical grout or other techniques to seal the offending fissure.

By August, 1966, after successful sealing of the fissure and unwatering of the mine and cementation of underground workings as development advanced, exploration revealed that there was less ore available than Searls had hoped for, not enough to yield significant profit after paying for the estimated high capital expenditures to put the property into production. At Ruby Hill, as at Tombstone and as at Gray Rock, the ore is still there, but so is the water.

Not all Newmont's activities during the Depression were devoted to exploration or to gold mining. As the record shows, Fred Searls was a highly competent geologist, possessing an almost uncanny sense of the right way of resolving a knotty geological problem. However, his restless energy and searching mind drove his interests into other fields where, for example, he developed a fondness for investigating new metallurgical techniques and mechanical novelties of all kinds. This fascination with metallurgical innovation caused Newmont to undertake research on production of elemental sulfur from sulfide minerals forty years before this idea was pursued by practically every copper mining and smelting company in the United States and Canada. In 1930, however, the objective was not to remove sulfur from smelter fume or power plant stack gas as it is today, but rather simply to develop an inexpensive means of producing sulfur from sulfide mineral concentrate. In December, 1929, Newmont had been approached by the Sulphide Corporation with the suggestion that Newmont finance laboratory and pilot plant work on a sulfur production process developed by E. W. Wescott.

The idea appealed to Searls, and he agreed to have Newmont finance laboratory work followed by construction of a pilot plant to cost an estimated $300,000. If the process proved feasible, it was intended to build a commercial plant with a capacity of about 300 tons per day of sulfide concentrate feed at an estimated cost of $1.5 million. Texas Gulf Sulphur joined in this venture shortly after Newmont decided to go ahead.

The reactions in the process were based upon release of sulfur from metal sulfides by means of a reaction with chlorine or a chloride. The sulfur was produced as a vapor and the chlorine was recirculated. Half the sulfur produced was to be re-used in the process, and the remainder was condensed for sale.

Work on the Sulphide Corporation's process continued sporadically all through the Depression, but in 1938, Newmont decided it had had enough. There were better opportunities for its funds elsewhere, and the pilot plant that had been built in Buffalo was shut down.

The main difficulties the pilot plant developed were that it appeared to be impossible to produce a clean sulfur, a great deal of dust being carried over with the sulfur vapor. Also, maintenance of heat and gas volume balances in the rather complex system became quite difficult. Finally, a great deal of sulfur dioxide, nearly 6 per cent by volume, was formed in the exit gas before the sulfur could be condensed. It appeared that this back reaction took place between sulfur as a vapor and the carbon dioxide present in the exit gas.

To be strictly accurate in reporting, nearly all the ventures described in this chapter thus far would have to be presented simultaneously. They happened almost all at once, or very closely spaced. One can only appreciate the changing Newmont management attitudes by stopping now and then to trace the development of new approaches. Thus, a major shift in the expression of the basic Newmont philosophy began in the middle 1930's. This arose because of a growing concern in the Newmont management over the proliferation of Federal government agencies directly involved with business. Raising the price of gold was one thing, but the Blue Eagle, the N.R.A., the death of the "little pigs," the Wagner Act, the S.E.C., all the measures Roosevelt took to alleviate the Depression and to establish what he considered necessary controls over business activities, simply served to deepen the average businessman's distrust and alarm.

Judge Ayer in particular thought that the United States was headed for destruction, and he was quite receptive to Fred Searls's suggestion that Newmont begin to devote more and more attention to mining outside the United States, particularly in Southern Africa and Northern Rhodesia. Furthermore, in the late 1930's mining costs had begun to climb and were beginning to cut into the profit margins of gold mining. Also, base-metal prices were definitely showing signs of life once more. Therefore, the focus of Newmont's attention began to shift from gold mining in the United States and in Canada back to base metal mining, both in the United States and elsewhere.

The opinion that the outlook for American industry was rather dim in the late 1930's, was held not only by some of the Newmont directors and members of its management but by the W. B. Thompson family as well. Thompson's daughter, Margaret, who had by then married Anthony J. Drexel Biddle, Jr., had become well acquainted with Ernest Oppenheimer, whom she found to be a most charming and most impressive man.

Those who knew Mrs. Biddle describe her as equally charming and impressive, especially in that she seemed to have inherited something of her father's business acumen. Throughout her adult life she maintained an active interest in the affairs of the company founded by her father. Living in Paris as she did, Mrs. Biddle's attractive personality and lively interest in business as well as in the arts brought her a wide circle of friends, which included such men as Oppenheimer.

To review briefly, Newmont in 1928 had taken a 25 per cent interest in Rhodesian Anglo American Limited, a subsidiary of Anglo American Corporation of South Africa formed by Ernest Oppenheimer. This commitment, as mentioned previously, proved somewhat embarrassing at the time because Newmont had to borrow money to pay for its Rhodesian Anglo American shares. In response to a Rho Anglo request, Newmont furnished operational know-how in the person of Harold S. Munroe, who was loaned by Newmont to direct Rho Anglo's technical staff with headquarters at Broken Hill in Northern Rhodesia. Munroe had worked at various properties in the West and in Mexico, and came to Newmont from American Metal Company in 1928.

Anglo American, the parent company of Rho Anglo, also

asked Newmont in 1939 to recommend a man experienced in copper metallurgy who could come to Rhodesia and assist in designing new smelters. Fred Searls had no hesitation in recommending Magma's McNab. In a letter dated August 12, 1929, to Carl Davis, consulting engineer for Anglo American, Fred Searls wrote: "We respectfully suggest that McNab has few equals and no superior in the business of copper metallurgy." A similarly strong recommendation for McNab was written by Henry Krumb. In November, 1929, McNab went to Africa, where he was most helpful and effective in plant design and in the solution of technical problems for N'Changa and Rhodesian Congo Border Concessions.

All this was fine on the surface, but it became apparent to Mrs. Biddle, through conversations with Ernest Oppenheimer sometime in 1933, that all was not well in Northern Rhodesia from the point of view of the effectiveness of Harold Munroe, of whom Oppenheimer did not seem to have a very high opinion. On the other hand, Mrs. Biddle's conversations with Munroe impressed her with his competence, and it occurred to her and to Fred Searls that one explanation for the seeming lack of rapport between Munroe and Oppenheimer could be that Munroe needed more of an official standing in Newmont.

At about the time this idea developed, Fred Searls and Franz Schneider were in London discussing various matters with the Anglo American management, including the formation of a new company or two in Rhodesia and South Africa. Schneider, concerned by what he thought might be an unwise development in Munroe's promotion, returned to New York by way of Paris, where he was entertained by the Biddles and was also exposed to Mrs. Biddle's forceful opinions regarding affairs in Rhodesia.

However, finding himself unable to support a promotion for Munroe, Schneider became quite perturbed at what could easily develop into a troublesome confrontation. Several other officers and board members were equally concerned, and a period of discussion and uncertainty developed. It was described by one member of the Newmont board in this way, "There has been entirely too much individuality and not enough cooperation in our organization to suit me. However, we can't always have our own way in everything. Just think how we have opposed Fred in some of his ideas." It was a new experience for Newmont, and a disturbing one.

Eventually, any possible controversy was avoided when Mrs. Biddle resigned from the Newmont board in 1934, and Harold Munroe was elected to take her place. Munroe continued with the Newmont board until 1936, when he resigned, and Carroll Searls, Fred Searls's brother, was elected a director to fill the vacancy.

As the years went by, however, Newmont seemed to grow further and further away from the center of events in Anglo American. Oppenheimer had a passion for floating new companies, either operating mining companies or those of the investment trust variety. Each time he did so, Rho Anglo would increase its capital, and somehow Newmont never seemed to grasp the opportunity to take its proportionate share of new stock. Newmont's interest in Rho Anglo thereby became progressively diluted. Nevertheless, Newmont retained an interest in Rho Anglo until 1947, with the exception of one block of stock that Fred Searls sold in order to provide funds in part payment of the purchase of O'okiep.

By 1947, Newmont's interest in Rho Anglo was down to about 5 per cent from its original 25 per cent. By then, Newmont had interests in Africa other than Rhodesian copper, and all concerned had become somewhat frustrated over the complexity of Rhodesian copper mining company ownership. Therefore, Fred Searls assigned Plato Malozemoff the responsibility of disposing of the remaining Rho Anglo shares. Because of currency regulations in effect at that time, this was not easy, but by various switches to Australian companies, Malozemoff was able to recover the investment, although the series of transactions was not completed until 1953.

Happier circumstances surrounded Newmont's return to O'okiep in South Africa. The price of copper was rising, and not even the so-called Roosevelt Depression of 1937, could discourage copper men. Consequently, it seemed a good time to reactivate the South African Copper Company, the owner of the O'okiep property in Namaqualand, South Africa, which had been formed back in 1928 by Newmont and American Metal Company. The Newmont-American Metal exploration program and the decision to suspend the O'okiep project in 1932 are described in Chapter 3.

Copper mining in Namaqualand goes back over one hundred years. The Cape Copper Mining Company was organized in January, 1863, and from its operations in Namaqualand, paid dividends totaling £1,324,375 from 1864 to 1888. Reorganized in 1888 as Cape Copper Company Limited, the property produced £3.5 million in dividends up to 1918. Major source of the dividends was the East O'okiep copper orebody, running about 20 per cent copper, from which over 1 million tons of ore were mined. From the Nababeep orebody, about 900,000 tons of ore were mined, running 4 to 5 per cent copper. All this was high grade enough to be shipped to Europe for smelting.

As the end of the easily accessible ore approached, Cape Copper Company, for some strange reason, looked everywhere for additional ore reserves except in the neighborhood of O'okiep. The company examined the Tilt Cove property in Newfoundland, and the Rakha Hills property in India, as well as many other properties. In general, these explorations were poorly conceived and highly expensive, with the result that the African mines had to close in 1919, and the Cape Copper Company Limited went into receivership that year.

At the end, the company did try desperately, but unsuccessfully, to find more ore in its own neighborhood. Having always mined only high-grade, direct-smelting ore, it apparently occurred to no one in the company that anything could be made of the two or three lower-grade orebodies that they did discover. Had they tried using concentration by flotation, they could have saved themselves, but they seem never to have even tried it, even though the flotation process was well known at the time.

It is reported, however, that a geologist named H. M. Kingsbury was employed by Cape Copper Company in the search for ore and that he advocated use of flotation to treat the lower-grade ore he found near O'okiep. If so, Cape Copper must have turned him down and marched blindly on into bankruptcy.

Kingsbury kept the faith despite this blow and tried to interest others in what he felt sure was a good property. About six years later, he and a friend, Augustus Locke, of leached out-crop fame, approached American Metal Company with the suggestion that Amco take over the Cape Copper Company's properties. In March, 1926, American Metal made an agreement with Kingsbury and Locke regarding such a deal and in June, 1926, took an option on the Cape Copper Company's Namaqualand mines at O'okiep.

As noted in Chapter 3, Otto Sussman of Amco invited Fred Searls to visit O'okiep in 1927, and the result was that in early 1928 American Metal and Newmont formed South African Copper Company, in equal shares, to hold the option on Cape Copper, which it took up in 1931.

Although Newmont eventually acquired 73 per cent of South African Copper Company by successive purchases of stock during the early 1930's, throughout most of the O'okiep exploration period a one-third interest was held by United Verde Extension through Fred Searls' friend, James S. Douglas, or "Rawhide Jimmy" as he was widely known. In July, 1928, Douglas bought a one-sixth share in South African Copper from each of Newmont's and American Metal's interests, making ownership a three-way equal split.

Rawhide Jimmy got his name at Nacozari in Mexico in 1899, where, as manager, he found that steel cable-support rollers lasted longer if reinforced with wrapped layers of wet rawhide. The nickname was in no way a comment on Douglas' personal characteristics. Quite the contrary.

Douglas began to build his fortune in 1912, when he and a partner acquired the claims covering the ground where Doug-

las believed he would find a faulted segment of the high-grade
United Verde copper orebody at Jerome, Arizona. It was there,
all right, and, as United Verde Extension, Douglas' discovery
paid a total of about $52 million in dividends before it closed
down in 1938.

When it came to O'okiep, however, Douglas was less per-
ceptive. Deciding to liquidate all his copper holdings, he sold
his entire interest in South African Copper Company to Newmont
in 1936 for a reported $30,000. Douglas thus missed amassing
a second fortune when O'okiep turned out to be an even better
dividend payer than UVX, not that Douglas really needed the
money.

In reactivating O'okiep in 1937, it appeared more appropriate to form a new company. The South African Copper Company was therefore liquidated early that year, and its assets were
sold to the newly formed O'okiep Copper Company, incorporated under the laws of what was then the Union of South Africa, with Henry Smith as president. By May, 1937, all the assets
of the South African Copper Company were transferred to O'okiep
Copper Company for $2,667,000. The present ownership of
O'okiep Copper Company is: Newmont, 57.5 per cent; Amax, 18
per cent; various financial institutions and the public, 24.5 per
cent. Of influence in the formation of O'okiep and the decision
to activate the project were certain special tax incentives
granted by the South African government to copper mines in
Namaqualand, where the O'okiep orebodies are located. These
incentives served as a model for a number of similar enterprises,
the second of which was the Palabora project that was organized
in 1956.

Because of his work at O'okiep during the exploration period before the Depression, George Kervin was clearly the man
to bring O'okiep into production. Therefore, in 1937, Kervin
left Nevada City where he had been manager of the Murchie-
Zeibright division of Empire Star, and went out to South Africa.
Hendricks came down from Northern Empire in Canada and replaced Kervin at Nevada City. To help him in Africa, Kervin

had the assistance of Wesley P. Goss and Eugene H. Tucker, chief electrician of the Murchie-Zeibright group. To Kervin, going to South Africa was like going home. To Goss, at least the landscape surrounding O'okiep seemed familiar because the terrain and the climate were so similar to that of Jerome, Arizona, where he had worked for a number of years. To Tucker, however, who was born and brought up in Nevada City, California, going to O'okiep must have been a little like going to the dark side of the moon.

Kervin and his staff had to begin from scratch. There had been no real mining at O'okiep since the old Cape Copper Company folded in 1919. Outside of a 30-inch gauge railroad that ran 90 miles to Port Nolloth on the African seacoast, nothing usable was available on the property except a few rather tacky buildings. There is not space here to detail everything that had to be done, but anyone at all familiar with the mining business knows that Kervin and his staff accomplished a near miracle in developing three mines for production, building a 1,500-ton-per-day mill, a copper smelter, and a townsite, all active by August, 1940.

The fact that throughout its life, O'okiep has been recognized widely in the mining industry as one of the lowest-cost and most efficient mining and smelting enterprises in the world, owes much to the original impetus given the project by George Kervin, and even more to the energy and initiative of the three managing directors who brought O'okiep into full production: Henry DeWitt Smith, M. D. Banghart, and currently, David O. Pearce.

The smoothness with which O'okiep Copper Company was established and operated in South Africa also owes much to Albert Livingstone, a Johannesburg attorney, who handled O'okiep's legal affairs from the very beginning. Livingstone became a director of O'okiep in May, 1937, and remained on the board until his death in February, 1970.

On one of his trips to Africa in the mid-1920's, Fred Searls met Livingstone during an attempt to acquire an interest in a

Rand gold property. Searls was quite impressed by Livingstone, and retained him to help in organizing O'okiep. A man of "tremendous judgement," according to Malozemoff, Livingstone served ably and loyally on the Tsumeb board as well from 1947 to 1970. In addition, he helped Newmont greatly in the strenuous negotiations that led to the formation of Palabora Mining Company in 1956.

Kervin, aided by W. E. Meals, supervised the design and construction of the O'okiep copper smelter, built at a cost of only $800,000. The designed capacity was 14,000 tons of blister copper per year, but in fiscal 1948, its seventh year of operation, the smelter turned out 21,359 tons of blister with no additional capacity provided. Production was steadily increased, passing 25,000 tons annually by 1954.

Thereafter, as new mines were opened up and mill capacities were increased, the reverberatory furnace had to be enlarged and a second converter added. Over the years, the O'okiep smelter staff has steadily improved its techniques, so that today the smelter has a capacity, without major enlargements, of about 50,000 tons of blister copper a year, although concentrate has never been available to permit reaching that output.

However, Kervin saw only the first year or so of the smelter's operation. In 1941, not feeling well, Kervin asked to return to the United States, and M. D. Banghart was sent out by Henry Smith to relieve him. The story of O'okiep thereafter is the story of the work, successively, of Messrs. Smith, Banghart, and Pearce, and the many men who worked with them in the extraordinary accomplishment of O'okiep.

In 1938, before production began, the O'okiep ore reserves stood at about 6.9 million tons at Nababeep, averaging 2.52 per cent copper; 2.7 million tons at East O'okiep, averaging 2.33 per cent; and 600,000 tons at Narrap, averaging 2.18 per cent; for a total of 10.2 million tons averaging 2.45 per cent. Under the guidance of Henry DeWitt Smith and Banghart, vigorous exploration was undertaken, so successfully in fact, that ten years later, the O'okiep sulfide ore reserves stood at 11,537,000 tons, averaging 2.67 per cent copper, even though in the intervening decade O'okiep had milled over 6 million tons, averaging 2.45 per cent copper, and had produced 138,822 tons of blister copper.

A capable administrator, M. D. Banghart established at O'okiep an excellent record for operating efficiency and cost control, especially during the difficult times of inflation after the end of World War II.

For his accomplishment at O'okiep, The American Institute of Mining Engineers in 1961 presented Banghart with the William Lawrence Saunders Gold Medal "For distinguished achievement in mining other than coal." The citation read in part: "For the development of new mining techniques which overcame many operating difficulties and resulted in O'okiep and Tsumeb's becoming low cost producers under his capable management."

When Banghart came to New York in January, 1954, to become Newmont's vice president for operations, he was succeeded as general manager at O'okiep by David O. Pearce. A native of London, England, Pearce went out to Africa when he was twenty-one years old. Obtaining a diploma in mining engineering from the University of the Witwatersrand, Pearce began his career with Anglo American Corporation of South Africa in one of their Rand gold mines. His work as surveyor and shift boss was interesting enough, but the deep mines of the Rand were "too bloody hot," and Pearce moved on up to Rhodesia, where he worked, successively, in the gold mines of the Rhodesian Corporation and Falcon Mines, Limited, and eventually became general manager of both companies. Also, while in Rhodesia, Pearce was for eight years a member of the Executive Committee of the Chamber of Mines of Rhodesia. In 1955, he was offered the position of superintendent of mines at O'okiep, which he found more to his liking.

At O'okiep, Pearce became assistant general manager in 1957, general manager in 1959, and a director of the company in

1961. Since 1968, he has been managing director of O'okiep and Tsumeb, and also vice president for operations and a director of Newmont, positions he assumed on Banghart's retirement.

By various means, such as redesigning the stoping methods and replacing the existing 2⅝-inch drills with 4½-inch machines, Pearce was able to increase O'okiep Copper Company's total average mine production from 100,000 metric tons to around 270,000 tons of ore per month, although about one-third of this increase resulted from bringing in the Carolusberg mine in 1963. Use of the larger drills yielded an increase of over 400 per cent in the footage drilled per machine shift. Also, the use of sophisticated, large-size loading and haulage equipment greatly increased mining efficiency. Production costs actually declined from 1957 through 1967, but thereafter inflation took hold, and costs have climbed inexorably to a level somewhat higher than that of 1957.

Probably the chief difficulty at O'okiep has been that mining has to be done in several relatively small orebodies spread over a large area instead of in one or more large orebodies. An indication of the scattered nature of operations is that the company has to maintain about 150 vehicles to get the staff around to the various mines. O'okiep also operates three concentrators and a copper leaching operation, and not only is the movement of staff complicated, but distribution of supplies and repair parts is quite an undertaking.

The O'okiep mining method, initiated by Banghart and developed and modified by Pearce after him, is a sublevel stoping method. Sublevels are from 40 to 70 feet apart. Stopes are opened by developing a slot across the area to be mined and blasting ore into the slot from long blast holes drilled by 4½-inch machines. Cycloned mill tailings fill these open stopes after stoping is completed, and pillars are then removed by sublevel caving or mass blasting. Stope blasts average 20,000 tons each, but as much as 450,000 tons have been broken in one blast. At the time of writing, O'okiep was planning a 750,000-ton blast.

At O'okiep, as at many other mines, a steady increase in efficiency is required just to keep ahead of a declining grade. For example, in 1959, employment at O'okiep totaled 4,394 men who produced 1,689,000 tons of ore. In 1971, there was no significant increase in manpower, but tonnage for the year had been increased to 3.1 million tons. This increase in labor and mining efficiency was, however, largely offset by a decline in grade of ore from 2.46 per cent to 1.28 per cent copper.

From the beginning of operations through 1971, O'okiep had mined and milled 52,149,000 tons of ore yielding 904,600 tons of blister copper, which it sold for a net income of $251,-294,905. Out of this income, dividends of $223,357,280 have been paid, of which Newmont itself received $127,221,114. Ore reserves, as of the end of 1971, were estimated at 27,615,000 tons, averaging 1.55 per cent copper.

To limit this discussion to production and financial data, would be to overlook what O'okiep has meant to Namaqualand. From 1919, when the old Cape Copper Company went bankrupt, the territory existed as an arid, poverty-stricken land where the inhabitants waged a daily, endless struggle against an inhospitable climate and an unproductive soil. It cannot be said that O'okiep has changed the climate, but it is true that, led by O'okiep, other mining companies and other industries have come to Namaqualand, and the people are no longer solely dependent on the meager yield of agriculture.

Newmont had the courage to develop O'okiep when the price of copper was running around $0.13 a pound, and both the country and the company have benefited thereby. Mining is now basic to the whole Namaqualand economy, and current exploration indicates that mining may play an even greater role in the future.

Aside from these facts, O'okiep has developed, directly and indirectly, a much more wholesome and active community life in Namaqualand than ever existed before. Springbok, O'okiep, Nababeep, Carolusberg, Concordia, all are thriving, modern communities because of O'okiep and its mines.

Originally, O'okiep's work force had to be drawn from the

colored population of Namaqualand (to use the South African
term), but it soon became apparent that, for the most part, the
low standard of living prevailing in the area had so debilitated
the inhabitants that they were just not up to the hard physical
work involved in underground mining. The government then
allowed workers to be brought at first from Nyasaland, but
more recently from the Transkei.

The O'okiep Company looks after the health and welfare of
all its employees, providing necessary medical and recreational
facilities for all. In particular, the black employees, numbering
about 2,000, receive, in addition to food, housing, and medical
care, such training and educational facilities as will enable them
to improve their skills and earning power. Company policy is to
encourage such self-improvement in the belief that this is the
best means of aiding real economic progress in South Africa. For
this and other reasons, O'okiep is widely recognized as a good
place to work. Until the South African government prevented it,
men came to work at O'okiep all the way from Nyasaland (now
Malawi) at their own expense. They still would, but the government now requires that O'okiep's black employees come only
from the Transkei in South Africa. There are no blacks native
to the O'okiep area, and although the company would prefer to
have black employees and their families settle permanently at
O'okiep, the government has not as yet permitted it.

O'okiep has continued to employ colored workers wherever their abilities and governmental and union regulations permit, and now about 1,800 colored employees are on the O'okiep
payroll. Every effort is made to improve the skills and earning
power of these employees. For example, when the new Rietberg
mine was opened recently, O'okiep, by negotiation with the
miners' union, was able to staff the mine almost entirely with
colored employees, who receive wages approaching the level paid
to Europeans in comparable jobs elsewhere. Unquestionably,
the standard of living of the colored people in Namaqualand has
been raised by the operations of O'okiep Copper Company.
However, O'okiep is still hampered, as are other companies in

South Africa, by political and social regulations that prevent full development of the potential of all its employees. But much progress has been made and more may be expected.

Although the foregoing record is impressive enough, O'okiep deserves an even greater recognition than the bare production and financial data would indicate. The description of O'okiep has been deliberately placed at the end of this chapter because in the Newmont history it marks the end of one era and the beginning of a period of great growth. In the decade from 1930 to 1940, Newmont was involved in many mining operations, but all of them quite small and with profits commensurate with their size. To compare Newmont's second decade with its first, instead of an aggregate income of about $50 million from 1921 through 1931, Newmont's total income from 1931 through 1941 equaled only $25 million. However, about three-quarters of this came from mining activities initiated by Newmont and only about one quarter from capital gains on sale of securities, a marked contrast to the first decade, when the bulk of Newmont's earnings represented capital gains. In other words, by 1941, Newmont had become a mining company in earnest, and its stock portfolio, grown from a market value of $9 million in 1932 to $42 million in 1941, was now a means to an end, not an end in itself.

With the development of O'okiep, Newmont saw its horizons widen, not geographically but economically. The income and the experience gained at O'okiep, strongly supplemented later by that at Tsumeb, gave Newmont the spirit and the funds to tackle larger and more complex mining enterprises. Here was a new opportunity for growth, and although World War II blocked progress for a time, Newmont in the postwar world took full advantage of its growing strength.

Idarado miners, 1945

6 War and Its Aftermath

The period of World War II represents something of a flat spot in Newmont's growth. All the company's energies were devoted to increasing production at existing properties, especially after December 7, 1941. Less attention was given to finding new orebodies, except for assistance given the Atomic Energy Commission in the search for uranium.

Certainly, the war put the finishing touches on Newmont's status as a gold miner. If the company had not already turned away from gold exploration, the war would have forced the move. The fact was that the $35.00 price of gold, even by 1937, was looking less exciting as costs rose toward it. The very price stability that in 1932 attracted Newmont so strongly had become in only five years somewhat of a deterrent. A miner has to be flexible and quick to adjust to this sort of thing.

As it was, in 1939 Newmont was still actively engaged in the management of twelve operating gold mines, from eight of which it received dividends. At the same time, Newmont had varying degrees of ownership in seven copper mining companies, but with the exception of O'okiep and Magma, it did not participate in the management of any of them, except in an advisory capacity.

Of course, in 1939 O'okiep was not in production, and Magma, as mentioned previously, was so independently managed as not really to be regarded as a Newmont company. Newmont's ownership of Magma was then less than 10 per cent.

The point is that for 1939, dividends from Newmont's gold mining properties totaled $656,661. On the other hand, Newmont in 1939 received a total of $1,564,810 in dividends from its copper mining investments. Magma's contribution was $178,200. All the hard work of managing was going into gold mining, but the dividends from copper were twice as great and at a fraction of the effort.

Thus, Newmont's decision in 1937 to reactivate the O'okiep project could not have been better timed. It gave Newmont a copper mine to manage, and it came into production just in time to help the war effort.

On the face of it, Newmont seems hardly to have noticed the outbreak of war. In the annual report for 1939 there is no mention whatsoever of the war or of any of its effects. Of greater investor interest, perhaps, was the fact that in 1939 Newmont became listed on the New York Stock Exchange. Prior to December, 1939, Newmont's common stock had been quoted on the Curb Exchange, which was Colonel Thompson's favorite area of operations. Newmont, in 1939, paid $3.50 per share in dividends on the 531,646 shares outstanding, approaching once more the $4.00 annual dividend rate established in the late 1920's on a slightly lesser number of shares.

Beginning in 1940, Fred Searls became engaged in assisting the work of the British Purchasing Commission in New York, and Bob Hendricks from Empire Star was actually employed by the commission. Other than selling the output of O'okiep to Britain, Newmont had little direct contact with the war effort, except for Henry Smith's being aboard the torpedoed *Athenia* in 1939 and being rescued from a lifeboat. The Japanese assault on Pearl Harbor in December, 1941, changed all that.

Within three weeks of the strike at Pearl Harbor, Newmont had signed a contract between the newly formed Metals Reserve Company and the Gray Eagle Copper Company, which was 98.9 per cent owned by Newmont. Gray Eagle Copper owned a small but high-grade orebody in northern California, which the old Mason Valley Company had purchased back in 1916.

Under the contract with Metals Reserve, Gray Eagle agreed to deliver 6,800 tons of copper in concentrates annually for a period of three years beginning in March, 1943. Newmont's engineers estimated that three years of mining at this rate would exhaust all the ore of commercial grade. To finance reopening the mine and equipping the property, Newmont put up $1 million. Before operations ended, the actual advances by Newmont totaled $1.8 million. When it shut down on June 23, 1945, Gray Eagle had produced about 14,680 tons of copper in concentrates, at a net profit of about $468,000 after all charges and repayment of debt.

When George Kervin returned from O'okiep in 1941, he was assigned as manager of both the Gray Eagle and the Resurrection properties, the latter in Colorado. Between Christmas and New Year's, 1941, Kervin and Bob Hendricks, who was back at the Empire Star, went up to Happy Camp in northern California to examine the Gray Eagle. Going through the old workings, the party came to a series of raises that led to an ore face the group wanted to see. Kervin got about halfway, and then said that he was too tired to continue. The rest of the party went on, but when they got back, found that Kervin had disappeared. They found him near the bottom of a raise, caught between two stulls. They rushed Kervin to a hospital in Yreka, but he had been so badly injured that he died within a matter of days.

In George Kervin's death, Newmont lost the contributions of an excellent mining and metallurgical engineer. A short, rather stocky man, and inclined to be uncommunicative, Kervin neither smoked nor drank, and as far as his associates recall, had no vices other than a passion for betting on horse races. Wesley Goss, who worked closely with Kervin at O'okiep in South Africa, thinks of him as one of the most competent engineers and all-around "good men" that he ever knew.

After Kervin's death, Gray Eagle was managed by Bob Hendricks until it was closed down in June, 1945. He was assisted by Wesley Goss, Hugh Steele, a geologist, who is now with Newmont Exploration, and Gene Tucker, who, back from Africa, was at Gray Eagle as mechanical superintendent until the property was closed down.

After December, 1941, demand for metals from the mines of North America grew beyond all precedent. The Metals Reserve Company, which had been set up by the War Production Board to encourage output of various metals needed for the war effort, found itself floundering and badly in need of an executive head. In early 1942, representatives of the WPB approached Fred Searls for help. He, in turn, persuaded Henry Smith to volunteer, and on March 17, 1942, Smith resigned from Newmont and became executive vice president of Metals Reserve Company. In this capacity, Smith had the entire responsibility for the agency then known as the Domestic Metals Procurement Program, which was made necessary by the great expansion in wartime demand for metals. In his work, Smith displayed an amazing ability to cut through red tape and to get things done without, at the same time, creating confusion. Such was his skill that he left no legacy of legal problems, nor did any of his contracts have to be renegotiated.

Fred Searls also took an active part in the war effort, although his connection with various Washington agencies was somewhat less formal than was Henry Smith's. Bob Searls, Fred Searls's son, recalls that in his opinion at the time, his father ran Washington during the week and Newmont over the weekend. Actually, Fred did do a great deal of commuting between New York and Washington, and he was to be found on most weekends at 14 Wall Street in the Newmont office. Fred became part

of a group known as "the Jimmy Byrnes Braintrust." In this capacity, with James Byrnes as the President's adviser in the
White House, Fred had considerable influence on more general
problems than the specific ones Henry Smith handled in Metals
Reserve.

Filed away in Fred Searls's encyclopedic mind were the sa-
lient points of just about all the mineral properties he ever visited, some of these dating back to his days as an exploration geologist for U.S. Smelting. One such property was the Black Bear
vein located at 12,000 feet in the Uncompahgre Range just west
of Ouray, Colorado. Discovered in 1894, the Black Bear had produced a fortune in silver, but as mining continued over the
years, the workings disclosed a zonal change wherein the zinc
content increased and silver declined with depth. Operated sporadically thereafter, the Black Bear mine was finally closed down
and abandoned in 1934.

Fred Searls first saw the property while he was still with
U.S. Smelting. He guessed, then, that there was a good deal
more ore at depth than the miners of the Black Bear had ever
dreamed of. Not until 1939, however, did Fred have an opportunity to begin to find out whether he was right.

A casual conversation with Oscar Johnson, president of the
Mine and Smelter Supply Company of Denver, along about
1938, reawakened Fred's interest in the Black Bear. Fred sensed
the growing demand for copper, lead, and zinc, and he felt sure
the Black Bear area had these metals in quantity, together with
additional credits in gold and silver. The area also had a proliferation of mining claims of scattered ownership, and only if
these were consolidated could a viable enterprise be established.
Another friend of Fred's, Spencer Hinsdale, a Portland banker,
became interested, and together with Oscar Johnson, they began
purchasing or leasing claims in the area. Finally, they approached Sunshine Mining Company of Wallace, Idaho, to see
whether Sunshine would be interested in a joint venture at the
Black Bear. Sunshine was much interested, and in 1939 the Idarado Mining Company was put together with Newmont, Sunshine,

Oscar Johnson, Hinsdale, and another friend of Fred's as participants.

It was Searls's intention to look for the downward extension of the Black Bear orebody by exploring upward from the Treasury Tunnel, which entered the mountain at an elevation of about 10,000 feet, some 1,060 feet below the lowest of the old Black Bear workings. The Tunnel portal is 2,000 feet above Ouray and south of that town. Working from the existing Treasury Tunnel, Idarado began a program of diamond drilling, drifting, and raising. In 1940, this exploration cut a vein on the Treasury Tunnel level, but it was the Barstow, another old-time producer, not the Black Bear. Production possibilities were not encouraging, and Idarado was therefore inactive throughout 1941 and 1942.

However, in 1943, the Metals Reserve Company saw an opportunity to increase production of copper, lead, and zinc by reactivating Idarado. In June, 1943, Idarado signed a contract with Metals Reserve whereby the latter leased the Idarado property with the objective of lengthening the Treasury Tunnel some 7,000 feet to find the extension of the Black Bear vein that Fred Searls still believed in, although he was reluctant to use Newmont's money to go after it just then.

To extend the tunnel, Idarado brought in Long John Austin, a well-known tunnel-driving contractor from Denver. Austin drove the tunnel 7,000 feet in an extraordinarily short time, and at a point 8,700 feet in from the portal, the tunnel hit the Black Bear vein. To check continuity, Searls had a drift run out into the hanging wall, from which four crosscuts were driven over to the vein, each one hitting it. However, the ore exposure was not exactly exciting, and Metals Reserve became quite discouraged and wanted to give up. By that time Searls's blood was up and he refused to be defeated. He ordered the driving of a raise 1,060 feet up from the Treasury Tunnel, to the old 6-Level of the Black Bear mine, and it was in ore all the way. With other raises and drilling, a sizable shoot of copper, lead, zinc, gold, and silver of ore grade was disclosed.

This made all the difference, and Searls promptly decided on production. There remained the questions of mill metallurgy and milling costs. Frank McQuiston, having become Newmont's chief metallurgical engineer, was working at Leadville, Colorado, on the Resurrection mine metallurgy. Searls arranged to have some samples of the Idarado ore sent to Frank with a request from Phil Kraft in New York that he find out whether the three base metals could be separated by flotation, and that he telephone Phil the results as soon as he had finished his tests.

McQuiston finished a few tests that indicated clearly that a three-product separation could be made by flotation with a reasonable recovery in separate concentrates of copper, lead, and zinc. He finished estimating the Idarado milling costs late one night, and in accordance with his instructions, McQuiston picked up the phone and called Phil Kraft at his apartment in New York. Phil finally answered the phone and listened patiently to McQuiston's account of the test work and its results. When McQuiston had finished, Phil said rather plaintively, "That's fine, Frank; just send me a copy of your report, and the next time, would you mind waiting until morning to call me?" McQuiston had overlooked the fact that he had called at 1:00 A.M. New York time.

With the promise of production made possible by finding the extension in depth of the Black Bear vein, Newmont and Sunshine bought back the Metals Reserve Company lease. In 1944, Newmont, itself, underwrote a new issue of 1,170,000 shares of Idarado stock at $1.00 per share. Although there was a mill on the Idarado property, built by San Juan Metals Company years ago, it was too dilapidated, even though operable, for even Fred Searls's taste for used equipment. Relucantly, therefore, Searls decided to have Idarado build a new mill to have a capacity of 250 tons of ore daily.

With McQuiston supervising design, Idarado began building the new mill in November, 1945, and despite a snowfall that winter of 450 inches, the construction was completed by the next summer. Fred Searls visited the property shortly thereafter, and the staff took him into the one and only stope that was entirely within sulfide mineralization. It looked like a jewel box. Actually, this stope was producing about all the mine's output, but somehow Fred got the impression that it was but a sample of a much greater reserve awaiting development. Searls promptly told McQuiston to increase the mill capacity to 1,500 tons per day and refused to be dissuaded. Among other things, McQuiston therefore ordered three Marcy 7 x 7 ball mills, which turned out to be surplus when the mine proved unable to increase production much beyond its existing rate of 250 tons per day. Two of these ball mills were sold later to Phelps Dodge for use in Bisbee, and the third mill went into storage, but is now operating at the Kombat mine of Tsumeb in South West Africa. For once, McQuiston had been allowed to buy new equipment, and then couldn't use it, after all!

Despite the discovery of new ore, Idarado still suffered difficulties. In 1945, the company milled 47,000 tons of ore, but wound up with a loss of about $32,000 for the year. Milling rate was stepped up to 66,000 tons in 1946, and Idarado showed a profit that year of about $55,000.

As time went on, a good deal of contention had arisen between Newmont and Sunshine, and finally the partners agreed to disagree. Mike Romney, a U.S. Smelting geologist from Salt Lake City, came in to appraise the property, and following his work, Fred Searls set a price of $1 million on the Idarado and told Sunshine that he was prepared either to buy or to sell at that figure. Sunshine sold, and Newmont then became the owner of 89.9 per cent of Idarado.

From the beginning of operations, Sunshine Mining Company had provided supervisory personnel for Idarado. John Edgar, mine superintendent from Sunshine in Idaho, was the general superintendent for Idarado. A most capable and engaging mining engineer, Johnny Edgar did a good job for Idarado. After returning to Idaho when Sunshine sold out, he advanced from mine superintendent to general superintendent of Sunshine and eventually to vice president. When ownership of the company changed hands some years ago, Edgar left Sunshine to join Bechtel in San Francisco.

After Edgar's return to Idaho, Fred Searls brought in his old friend, Billy Plumb, from California, as manager. However, Plumb suffered a heart attack in 1946 and was replaced by Fred Wise, who came to Idarado from the Getchell mine by way of the navy. Many of his associates credit Fred Wise with having made Idarado into a really good mining property. He continued the program of consolidation of neighboring claims, later carried on by his successor, John Wise, an associate but not a relation of Fred's. Fred Wise died in 1952, a great loss to the company.

John S. Wise came to Idarado as chief engineer in 1943 through his acquaintance with Johnny Edgar. A native New Yorker, as was Phil Kraft, John Wise was educated at White Plains High School and at Columbia University College of Mines. Following graduation from Columbia, Wise went to work for Sunshine at Wallace, Idaho, in 1938, where he served as mine engineer, miner, and shaft sinker. At the outbreak of war, Wise entered the Corps of Engineers of the U.S. Army, but at the request of Metals Reserve, he was released in 1943 to go with Idarado as engineer.

Wise served as chief engineer for Idarado until September, 1951, when he became mine superintendent. Appointed acting manager of Idarado in April, 1952, he became general manager in September of that year. For ten years thereafter, John Wise served as general manager, becoming president of Idarado and a director in 1963. However, in October, 1962, Wise became a mining engineer on the staff of Newmont, headquartered in New York City, and in January, 1965, he was assigned to Magma Copper Company at San Manuel, Arizona, as assistant to the president.

Wise has worked at San Manuel ever since, as assistant general manager, acting general manager, general manager, and in April, 1969, as vice president of Magma Copper Company. He is now also a director of Magma.

Eugene H. Tucker also played an active part during the

early days at Idarado, where he assisted the effort, in cooperation with the U.S. Bureau of Mines and Joe Ruth, of Denver, to develop the use of diesel haulage in the Treasury Tunnel. Thanks to certain improvements in equipment, including an efficient exhaust gas scrubber, Idarado became among the first underground mines to make consistent and safe use of diesel haulage underground. Idarado even built a complete 17-ton diesel locomotive in its own shops.

In 1951, Gene Tucker moved his family back to Grass Valley and made his headquarters there for several years. During that time, he traveled to Newmont properties all over the world. In fact, it would be quite difficult to name a property in which Newmont has an interest that Gene Tucker has not visited, or to which Tucker has not contributed assistance or ideas. In 1954, he was even loaned to St. Joseph Lead Company to work on a new mill the company was designing at Indian Creek, Missouri.

Following about a year of work in connection with the first nickel pressure leaching plant in the world, developed by Sherritt Gordon in Canada, Tucker finally made his headquarters in the New York office of Newmont in 1955, where he has been ever since.

Frank McQuiston has many metallurgical accomplishments to his credit, but one of the most outstanding is the copper-lead-zinc separation he worked out at Idarado in 1944. Still in use today, the flow sheet uses cyanide for depressing chalcopyrite and thereby separating it from galena. This technique was not new at the time, but McQuiston worked out a variation in reagent combination that yielded an excellent separation of the copper, the lead, and the zinc minerals into high-grade concentrates. A danger in the use of cyanide at Idarado was that it could dissolve considerable free gold. At first McQuiston tried using activated carbon and zinc shavings in boxes as a means of preventing gold loss, but eventually a method was developed which, by forming a chemical complex between the cyanide ion and zinc and calcium salts, permitted an effective copper-lead separation but prevented gold dissolution.

What happened to Idarado in the post-war world is another story. It appears in Chapter 13.

A most interesting, although thus far not very profitable, venture for Newmont in Colorado is the Resurrection mine in the hill above Leadville to the east. Newmont's interest in Resurrection came about through a conversation between Fred Searls and an old friend of his, R. T. Walker, a consulting geologist formerly employed by U.S. Smelting, Refining and Mining Company. Walker had long been convinced that oxide gold ore could be found in the neighborhood of the No. 2 shaft of Resurrection at the foot of Mosquito Pass. Neither Walker nor Searls was able to interest U.S. Smelting in Resurrection, but through his friendship with the Hecla management, Searls did get the Hecla Mining Company of Wallace, Idaho, to come in for 50 per cent of the venture Searls planned for the Resurrection. When the reorganized Resurrection Mining Company began work in 1938, Hecla managed the enterprise.

Byron Wilson of Hecla came down from Wallace in January, 1939, to manage Resurrection. He was joined three months later by Henry Volkman, also of Hecla, who came to Leadville to handle the Resurrection accounting. Volkman liked what he saw of Newmont, and in August, 1940, he went on the Newmont payroll, where he has remained ever since. Stationed at Leadville until July 4, 1948, Volkman came to the New York office of Newmont, where he now is responsible for all the company's purchasing activities and all insurance policies and programs.

During 1939, Resurrection did enough work to demonstrate that the oxidized gold ore Walker had hoped to find did not exist in payable quantitites. In addition, the company drove a drift 840 feet to connect with the Yak Tunnel, an old working 19,600 feet long that served only the useful purpose of draining some of the shafts of the district. Actually, it would take many more pages to recount in detail all the work of property acquisition, tunnel driving, and exploration that Newmont did around Resurrection at that time.

Despite the disappointment on the gold ore, enough evidence had been found of occurrences of lead-zinc-silver ore to enable Walker to convince U.S. Smelting that the company should join in Resurrection for one third of the venture. Ownership of Resurrection was thus split three ways equally, but thereafter Newmont was responsible for management of the mine.

Within a year, enough ore had been found to justify equipping a 250-ton-per-day flotation mill. According to Frank McQuiston, this was done with "junk" from other Newmont operations all over the West. With three different makes and sizes of ball mills and eleven different types of flotation machines, the mill presented a real puzzle to an operator. Most mining men could not even name eleven different types of flotation machines, let alone operate a mill equipped with that many.

McQuiston goes on to say that this equipment hodgepodge did not really affect metallurgy adversely, because results would have been bad in any circumstance, owing to the presence of marmatite, a zinc sulfide, that carried 10 to 20 per cent iron, plus a good deal of galena, oxidized and intergrown with the marmatite. Fortunately, nearby in Leadville, American Smelting's smelter was hungry for concentrate and took the Resurrection lead concentrate despite the poor quality.

Leadville in 1942 approached being a reincarnation of the Leadville of 1878. What with the Climax mine going full blast, and 25,000 ski troops in training at Camp Hale, both within short driving distance, the saloons, gambling joints, and the old red-light district were wide open. During that period, the army released forty volunteers to go to work for Resurrection in Leadville. McQuiston recalls that after two weeks, ten of these men were in jail, twelve had gone AWOL, ten had been fired, and the remaining eight turned out to be fairly good workers.

Anyone who visited Leadville in the early 1940's will surely never forget the Vendome Hotel, with its coiled rope fire escapes in each bedroom, its maids in evening gowns turning down the beds, and its floors that sagged dispiritedly as one walked toward the rear of the hotel. For his first few nights in Leadville, McQuiston paid the Vendome 50 cents a night to sleep on the lobby floor. He was promoted quickly, however, to a lobby chair at $1.00 a night, and eventually to an upstairs room at whatever the hotel thought it could get. A happier experience was dining at a restaurant down the street called the Golden Burro, affectionately referred to as the "Brass Ass." The proprietor, Charlie Frey, is said to have won the Brass Ass in a poker game, but he did build it into a restaurant with a terrific reputation for the quality of its steaks and salads. There were times in 1942 and thereafter when the Golden Burro served an average of 900 meals a day with a seating capacity of only thirty.

In 1942, Resurrection milled 75,000 tons of ore for a loss in the year of about $30,000. This loss was caused in part by the company's having purchased the Yak Tunnel and spent a good deal of money rehabilitating it. However, with the discovery of the Fortune orebody, prospects for Resurrection seemed brighter. The next year Resurrection qualified for receipt of incentive payments under the wartime Premium Price Plan, which was operated by the Federal government as a means of encouraging production of strategic metals by mining companies that could not have operated profitably without this assistance.

After the war, Newmont explored and operated Resurrection sporadically. After a long shutdown, the property was again reactivated in 1969 and entered production in 1971, with Newmont and American Smelting and Refining Company in equal partnership in a venture described in Chapter 13.

Demand for copper, lead, and zinc showed little sign of slackening after the war. Not until 1948 was there a substantial turndown in metal prices, but with the advent of the Korean conflict, demand for metals increased and prices strengthened again. Based largely on its experience with wartime shortages of base metals, the government of the United States determined to build up its strategic stockpiles of metals and a whole list of other materials. To accomplish this and, at the same time, to aid the economies of practically all the countries in the rest of the world, the Federal government set up the Economic Cooperation Administration, one division of which was the Strategic Materials Division, headed by Evan Just, formerly chief editor of *Engineering and Mining Journal.*

Through an odd combination of circumstances, Evan Just, Mrs. Biddle in Paris, and Jean Walter, a French architect, also in Paris, along with Frank Cameron, of St. Joe, and Fred Searls, became involved in a lead-zinc mining enterprise in northern Africa. The venture began years before World War II when Jean Walter had received a mineral property in Morocco in part payment of a large fee for some architectural work he had done. Then when his son Jacques had completed his university studies as a geologist and mining engineer, Walter had sent him to Morocco to examine his new property. Even to a neophite geologist, the possibilities were quite evident; samples taken on the property turned out to assay 18 per cent lead and 8 per cent zinc. Accordingly, Walter formed the Zellidja Mining Company, with himself as chairman and Jean Lacaze, his wife's brother and his comrade in the French Resistance during the war, as managing director.

Although the Zellidja Mining Company had become a reasonably profitable operation, it became clear after the war that if Zellidja were really to prosper, it needed more money and more technical help than Walter and Lacaze could supply on their own. In a repetition of history, Walter followed Oppenheimer and others to J. P. Morgan, and was referred to Newmont.

Fred Searls had a look at Zellidja in 1946 and liked it. Quite early in his investigation, however, he found that Frank Cameron, an old friend who was then vice president of St. Joe, was hot on the same trail. Swallowing whatever chagrin each felt, the two friends agreed that cooperation would be more productive than combat.

Fred's study of Zellidja also brought Jean Walter into contact with Mrs. Biddle, Colonel Thompson's daughter, who had returned to Paris after the war. Mrs. Biddle was enchanted with Mr. and Mrs. Walter and Jean Lacaze and thereafter devoted much of her considerable talent to furthering cooperation between Newmont, St. Joe, and Zellidja. Fred Searls and Mrs. Biddle evidently sparked enthusiasm from each other, with the result that the North African operation received for years more attention and staff time than it really deserved in view of the property's final financial outcome.

Following lengthy study of the Zellidja area in Morocco by Newmont and St. Joe geologists, including both Searls and Cameron, the partners proceeded in 1946 to negotiate a deal with Jean Walter for acquiring mineral rights in Morocco, or possibly an interest in Zellidja, itself.

In this negotiation, Searls was up against one of the shrewdest bargainers he ever encountered in the person of Jean Walter, who was backed up by an extraordinarily competent French geologist, Jacques Ségaud. The discussion led to an agreement to form a new company, Nord Africaine du Plomb, or NAP for short, in which Zellidja would have a 51 per cent interest and Newmont-St. Joe, 49 per cent. The 49 per cent interest was divided up 65 per cent Newmont and 35 per cent St. Joe.

Then the negotiation turned to which areas, and how much of each, should become the property of NAP. Searls, and his friend R. T. Walker, the consulting geologist, had become convinced that the Zellidja lead orebodies had originated in mineral-bearing solutions coming from south of Zellidja. They held this view in spite of Walter's belief, based on Ségaud's opinion, that the trend of the Zellidja mineralization was to the east. In fact, Walter had opened the talks by offering to assign NAP the available mineral properties in this eastern area.

Fred demurred, asking for extension of the NAP holdings to the south, considerably beyond the original boundaries suggested by Walter. In time, Walter gave in, but he did insist on an exchange whereby Zellidja picked up more land to the east in return for NAP's increased holdings in the south.

As it turned out, Searls and Walker had, for once, bet on the wrong horse. Exploration to the south yielded nothing. In the area to the east, which NAP had given up to Walter, Zellidja found about 12 million tons of high-grade lead ore. NAP had to explore eastward beyond this area clear on into Algeria, but then did find a large and high-grade zinc orebody. To acquire it, NAP and Zellidja had to form a new company, Société Algérienne du Zinc (ALZI) in which ownership was the same as in NAP.

This involvement in Algeria, harmless as it seemed at first, led eventually to enormous difficulties when revolution developed in Algeria, beginning in 1955. If NAP had simply accepted Walter's first suggestion, Newmont and St. Joe might have made a profit several times as great as the $4 million to $5 million they eventually realized out of ALZI.

However, in 1948, all this was far in the future, and in the annual report for that year Newmont reported that "the NAP (including ALZI) holdings are now known to contain one of the major ore deposits in Africa."

Because the Zellidja mill could treat only lead ore, and the ALZI ore was high-grade zinc (12–20 per cent), a new mill had to be built and mine development started. At that point, Evan Just learned of the development, interested as he was in opportunities for his Strategic Materials Division of E.C.A. Through Fred Searls, Evan met Mrs. Biddle in Paris and quickly caught her enthusiasm for North African mining in general and the ALZI project in particular. Mrs. Biddle introduced Evan to Jean Walter and Jean Lacaze, who were managing NAP and ALZI as well as Zellidja, and the result was an E.C.A. loan of $4 million, to be used for prospecting in the NAP area and to expand the Zellidja mill to a capacity of 4,000 tons per day. The loan was repayable by future delivery of lead and zinc to the Bureau of Federal Supply.

Mine development in the NAP-ALZI area called for sinking two 1000-foot shafts, one on the Moroccan side of the border and the other, the El Abed shaft, off in Algeria at the presumed eastern extension (again Fred's geological gamble, which turned out right this time) of the known ore zone. St. Joe sent over Kremer Bain, the company's shaft-sinking specialist. Not for the first time in Newmont's experience, heavy flows of water held up progress in the El Abed shaft for months, but by 1951, a large zinc orebody had been outlined in the ALZI holdings in Algeria adjoining the Moroccan border.

To supervise the design of the expanded Zellidja mill, Newmont assigned Frank W. McQuiston and Gene Tucker. General Engineering Company of Toronto did the engineering and construction. Zellidja's new mill was completed in 1951 and was inaugurated in a ceremony unmatched in the Newmont experience for rather bizarre splendor.

Present at Oujda for the inauguration were not only Newmont, St. Joe, and Zellidja executives and board members, but Moroccan and French government officials, a sprinkling of royalty, including the Duke and Duchess of Windsor, and General Juin, the only five-star general in the French Army at that time other than General De Gaulle. Crack Moroccan cavalry and French spahis went through maneuvers at full gallop. It all fitted perfectly the Moroccan background.

However, all this magnificence nearly ended in embarrassment when shortly before the ceremonies were to begin, Gene Tucker discovered that the primary crusher refused to work. On investigation, Tucker found that the manufacturer of the crusher had supplied an 18-inch main shaft when a 21-inch shaft had been specified. Together, McQuiston and Tucker hurriedly improvised means of supplying ore fine enough so that the primary crusher had nothing to do, yet there was a good flow of ore on the various conveyor belts, and the froth in the flotation cells was copious, albeit carrying nothing.

All the important visitors, except the initiated, went away believing they had seen the new mill in full operation. Both McQuiston and Tucker were decorated with the Order of Ouissam Alaouite Cherifien for "outstanding services to the Moroccan Government." With the crusher incident in mind, McQuiston and Tucker struggled fiercely to keep straight faces while General Juin kissed them on both cheeks in the course of the ceremony.

Despite a drop in metal prices after mining began in the

new zone in 1951, the E.C.A. loan had been repaid in metal by 1954 and the prospects for the North African enterprises looked excellent. There was only one sore spot. Zellidja, of course, had the richer ore reserves and the most profitable operation. Try as they would, Searls and Cameron could never pry out of Jean Walter an interest greater than 3.5 per cent in Zellidja itself. This fraction was divided as were the American holdings in NAP and ALZI, 65 per cent Newmont and 35 per cent St. Joe. To obtain even this minute portion required lengthy bargaining and a complicated process of stock exchanges. This transaction was only the beginning of many enormously complex tax, legal, and political problems in which Newmont and St. Joe found themselves involved.

Quite instrumental in solving some of these problems was Jacques L. Leroy, who has been with Newmont since 1955. Born in Belgium, Leroy had just turned fourteen when the "phony war" in western Europe suddenly became all too real. Jacques, his brothers and sisters, and his mother escaped from the Continent off the beach in Dunkirk in one of the last boats to get away under the German assault.

After a period in England, the family came to the United States, where Jacques was placed in a school in New Rochelle. Three years later, Jacques was back in England to enlist in the British Airborne Special Services Troops, which was a group conducting various missions behind the lines in eastern France and Belgium. During the remainder of the war, Leroy helped blow up bridges, wreck trains, and do counter-intelligence work generally.

After the end of the war, Leroy returned to the United States, entered college, and went through New York Law School. Thereafter, for five years, he was employed as an attorney with Davis Polk & Wardwell, a New York law firm. Through a friend, he became acquainted with Newmont, and it seemed to Malozemoff, who had become president of Newmont in 1954, that Leroy's special talents could be extremely useful, not only to Newmont generally but to the operations in North Africa in particular. Leroy was hired in 1955 and has been with Newmont ever since.

Along with the tax and other problems that Leroy was unraveling, there were certain inherent differences between the Zellidja mines and the NAP and ALZI holdings that became more apparent as time went on.

The Zellidja orebody was quite similar to the lead orebodies in southeast Missouri. Stope backs were competent, and ore lenses not so thick as to prevent room-and-pillar mining—a cheap method made even cheaper by using shuttle cars as coal mines do. The ore was high grade, 4–6 per cent lead.

On the contrary, the ALZI zinc ore, though high grade, averaging 18 per cent zinc in place, occurred in structurally weak rock. The stope backs had a habit of caving in. Consequently, Zellidja's cheap mining methods were out of the question, and the ALZI staff had great difficulty trying to mine the ore cleanly. An induced caving method was developed by John Kostuik, now president of Denison Mines, but it was quite expensive and resulted in a good deal of dilution. Ore that ran 18 per cent zinc in place, ran only 8 to 12 per cent in the mill ore bins. However, it was the only method that worked, but the dilution and the high costs cut deeply into ALZI's profits.

The Zellidja orebody, including that portion to the east that Walter traded away from NAP probably yielded between 35 and 50 million tons of ore before it had to close. The ALZI orebody never turned out more than about 12 to 15 million tons.

Actually, what Newmont referred to in 1948 as "one of the major ore deposits in Africa" turned out to be quite a bit less impressive than it had seemed. Further exploration to the east and north was disappointing, although an area to the east was eventually outlined containing ore running 3 to 5 per cent zinc in rock strong enough to permit low-cost mining. By then, however, time had run out.

Beginning in 1955, rebel activity in Algeria forced a curtailment of exploration and, in a few years, began to interfere with production as well. Not even Zellidja, well within Morocco, was safe from the Algerian guerrillas. In 1961, Zellidja and NAP were forced to curtail dividend payments, and in 1963, no dividends were paid at all.

By 1964, Newmont and St. Joe had given up on North Africa. For one thing, Morocco had obtained a substantial ownership of Zellidja by buying the stock inherited by Jacques Walter on the death of his father in 1957. For another, when Algeria gained independence in July, 1962, it promptly expropriated the ALZI mining properties. Not until April, 1971, did negotiations, ably led by Leroy, succeed in obtaining some indemnification for the mineral properties and also for Newmont's Algerian oil holdings which had been seized in November, 1970. It was a bitter end to an exciting, if not overwhelmingly profitable, venture into the northern end of the African continent.

There is one more Newmont venture, Tsumeb, which surfaced in the aftermath of World War II and demands attention. The property of Tsumeb Corporation Limited has turned out to be so extraordinarily profitable, and has had such a profound effect on the Newmont fortunes, that it seems appropriate and necessary to devote an entire chapter to this venture of Newmont into the somewhat mysterious and highly controversial area of South West Africa.

Tapping a Tsumeb lead furnace

7 Tsumeb — the Unique Orebody

From the discovery of its outcrop to its present-day productivity, the Tsumeb orebody has inspired such a string of superlatives that anyone not familiar with the property may be pardoned a certain incredulity. Nevertheless, it is true that few orebodies anywhere in the world have been as high grade in as many metals, have contained so many different minerals, have presented so complex a metallurgical problem, yet have proved as profitable as has the Tsumeb orebody.

However, aside from its geological and metallurgical interest, Tsumeb has been one of the strongest economic forces that projected Newmont into the front rank of the world's mining industry. Incorporated in South West Africa in 1947, Tsumeb Corporation Limited, in its first year reported sales with a value of $5.3 million at a net profit of about an even million dollars. Ten years later, for 1957, Tsumeb reported sales of about $52 million and made a profit of around $19 million.

Although Newmont's share in this phenomenal growth was just under 30 per cent, the effect of dividends from Tsumeb, and also from O'okiep, profoundly altered Newmont's postwar economic position.

Prior to the O'okiep-Tsumeb era, Newmont's dividend income had been running about $1.5 million to $2 million annually from domestic sources, and about $500,000 to $750,000 from foreign sources. With O'okiep and later Tsumeb in production, foreign dividend income began to climb, and in 1951, foreign dividends exceeded domestic dividends for the first time. Furthermore, total dividends to Newmont, which in 1945 were about $2 million, climbed to about $10 million in 1951, thanks largely to O'okiep and Tsumeb.

An income of this kind, increasing still more as it did in later years, enabled Newmont to think in much larger terms than the many small gold-mining ventures it initiated in the thirties. Furthermore, the success of O'okiep and Tsumeb marks the culmination for Newmont of emphasis on foreign exploration, a development that began, probably, with Judge Ayer's bitterness back in the mid-thirties over the various domestic reform regulations of the Roosevelt administration. Although it was well understood by the Newmont management, this policy was not expressed publicly until the annual report of 1948, which referred to Newmont's interest in foreign exploration as having been aroused by labor shortages and rising costs of operation in the United States. (It might be mentioned here, that Newmont's policy is now exactly the reverse. In 1968, for the first time since 1951, Newmont's net income from North American sources grew beyond its foreign net income, and in 1971, North American net income had climbed to 78 per cent of total net income.)

Considering its impact on Newmont's fortunes, one might expect the Tsumeb operation to be large and impressive. Impressive it is, but for reasons that are not immediately apparent. For one thing, Tsumeb is not easy to reach, except by air. It is 335 miles by rail or highway northeast of Walvis Bay, a port on the west coast of South West Africa. The country itself is about as large as Texas and New York combined, and although some of the land is arable and fertile, a high proportion of desert in and around South West Africa gives it a most forbidding aspect to the newcomer. Bounded on the east by the Kalahari Desert, on the south by the Namaqualand Desert, and along its western coastline by the Namib Desert, South West Africa was not, until fairly recent times, of any great interest to either the migrant or the tourist. Only from Portuguese Angola to the north was there, over the past century, any great migration of native tribesmen into the fairly fertile north central area of South West Africa.

Nevertheless, several distinct groups make up South West Africa's present native population of around 500,000. Most notable of these groups are the Herero and the Ovambo. A proud and intelligent people, the Herero, although nomads by choice, were skilled cattle raisers long before Germany annexed South West Africa in 1884. Still essentially nomadic and still cattle conscious, the Herero men have shown no interest whatever in mining as an occupation.

On the other hand, the Ovambo people, which make up about two thirds of the native population, live largely in native reservations, and are strictly responsible to their own governing body, the Owambo Legislative Council. Industrious and quick to learn, the Ovambo constitute the basic South West African work force.

South West Africa continued under German rule until the First World War, when South Africa took over the territory under a mandate of the League of Nations. Since the Second World War, neither General Jan Christiaan Smuts nor any succeeding nationalistic prime minister in the Republic of South Africa has been willing to recognize the authority of the United Nations to take over the League of Nations's mandate. At this writing, the United Nations's attempts to place South West Africa under the U.N. Trusteeship, or to establish the territory as an independent nation, have been unsuccessful and South Africa remains in control.

That the rigidity of this control is relaxing somewhat is evinced by the government's reaction to a country-wide strike of the Ovambo that swept South West Africa in mid-December, 1971. In a protest against the contract system under which they had been employed, some 13,000 Ovambo of the approximately 40,000 employed in the territory, refused to work and asked to be returned to their homes in Owambo near the Angola border.

Production at Tsumeb was reduced by over two thirds, and the management joined other employers in suggesting revisions in the contract system to provide mutually satisfactory employment arrangements. Following discussions among the government, the Owambo leaders, and the employers, a new system of employee-employer agreements was thrashed out. The old recruiting agency (SWANLA) was abolished, and the Owambo Legislative Council opened and operated new employment offices. Terms of the agreements were greatly liberalized, apparently to the Ovambos' satisfaction, for, after January 25, 1972, Tsumeb gradually built up the full complement of workers necessary, with some 50 per cent of former workers returning.

The history of the Tsumeb mining operation is well presented in a pamphlet called "Tsumeb, A Historical Sketch," written by Dr. G. Söhnge, published by the South West Africa Scientific Society in 1967 in Windhoek, the capital of the country. Dr. Söhnge, at one time chief geologist of O'okiep Copper Company Limited and later of Tsumeb, is now Professor of Geology at the University of Stellenbosch in South Africa.

Professor Söhnge points out that the word "Tsumeb" in the native dialect means "green slope." However, the name does not refer to a grassy hillside as one might expect, but rather to a rocky outcrop, on the slope of a hill, that was colored green because of the copper oxide content of the rock.

Mining activity in and around Tsumeb goes back two or three hundred years, but there was no effort at organized mining at Tsumeb until about 1885, when a trader and elephant hunter named Robert Lewis got a concession to the property from a chief of the Herero.

Before Lewis could even begin mining, the Germans, having annexed South West Africa the year before, wiped out his claim and took over his operations under mining rights granted by Chancellor Bismarck to the newly formed South West Africa Company. This company also obtained the right to build a railroad to serve the Territory, and a harbor at Swakopmund. Having evicted Lewis, the company began mining at Tsumeb under the direction of an Englishman named Matthew Rogers, who started sinking three shafts near the outcrop and did the first real exploration at Tsumeb.

When Rogers saw the Tsumeb outcrop in 1893, it stood up as high as 40 feet above the gradually sloping hillside. Composed mainly of quartz, the rock was shot through with green and blue copper oxides and silicates. Here and there were fissures of solid galena cutting through the copper minerals. Rogers wrote: "I very much doubt if I shall ever see another such outcrop." Rogers never did, nor is it likely that any one else will ever see such an outcrop, for there is no part of the world, except possibly in the Arctic or the Antarctic, where such an outcrop could by this time still be unobserved. Nor could the Tsumeb outcrop be spared for posterity, for it disappeared long since into a glory hole over the Tsumeb orebody.

In 1900, being in need of additional capital, the concession holders organized a new company called Otavi Minen und Eisenbahn Gesellschaft, to develop the Tsumeb mine and build the railroad from Swakopmund. The new company sank two new shafts near the outcrop, and promptly ran into a solid body of ore. Crosscuts from these shafts cut into almost solid galena (lead sulfide) containing veins of copper sulfide. Rapidly there was developed over 200,000 tons of ore running 12 per cent copper and 25 per cent lead. The low-grade ore ran 3 per cent copper and 4 per cent lead. The railroad to Swakopmund, on which this ore was sent on its way to European smelters, was completed in 1906. Except for brief interruptions during the First World War and during the Depression, the Otavi Company worked the Tsumeb mine continuously until the South African government shut it down by order during the Second World War. Total production at Tsumeb under German administration was 330,000 tons of lead and 180,000 tons of copper, most of it in the form of direct shipping ore. The Germans employed gravity separation of fine sizes of ore, and also built a copper-lead blast furnace as early as 1907. Eventually Otavi built and operated three blast furnaces, which, at their peak in 1927, produced 14,727 tons of copper black and 5,004 tons of lead.

The operation was quite successful financially. In the eighteen years between 1922 and 1939, the Otavi Company earned some $6 million and paid $5 million of that in dividends.

Arthur Storke, an American geologist employed by Selection Trust, a British company, had visited Tsumeb while the Germans were still operating the property before World War II. Also, A. F. Duggleby, on behalf of Newmont, paid Tsumeb a short visit in the 1920's. Neither visit led to anything, but during the Second World War, Storke went back to Tsumeb to see whether or not it could be reactivated for its possible contribution to the war effort. By that time, Tsumeb had become the property of the South African Custodian of Enemy Property, but Storke concluded that the problems of unwatering the mine, plus the obvious metallurgical difficulties, would require too much time and effort to make the mine immediately useful. Storke did, however, keep Tsumeb in mind as an attractive possibility for the future. After the war, Storke mentioned his ideas to Heath Steele of the American Metal Company. Steele told Storke that because of Newmont's position at O'okiep, American Metal would go in on any proposition at Tsumeb if Newmont would agree to run the property.

Having worked with Newmont on the original O'okiep exploration, Storke was well acquainted with Newmont personnel and therefore came at once to Henry Smith and A. J. McNab with his suggestion regarding Tsumeb. Thinking well of the idea, Messrs. Smith, Banghart, Söhnge, and E. N. Pennebaker, a consulting geologist for O'okiep, then visited Tsumeb in 1946, making the trip in a rather hair-raising ride by automobile some 900 miles over a rough track from O'okiep. Because the mine was flooded, the visitors had to content themselves with taking large samples of dump ore, which they sent back to the United States for metallurgical test work. Evaluation of Tsumeb reserves, based on the old German mine maps, was one of the early assignments Plato Malozemoff was given after he joined Newmont.

The samples and measurements taken by Smith and Banghart did, however, indicate that in the ore dump there existed about 500,000 tons of ore containing about 5 per cent copper, 16 per cent lead, and 11 per cent zinc. Examination of the records indicated that there should be available underground without further exploration approximately 670,000 tons of ore running 8 per cent copper, 21 per cent lead, and 12 per cent zinc. Metallurgical test work done by The Galigher Company in Salt Lake City, the American Cyanamid Company in Stamford, Connecticut, and by Dave Christie at O'okiep, showed that the dump ore, even though it was heavily oxidized, could be concentrated by flotation, or even by gravity concentration, to produce marketable products.

As a result of these studies, Newmont, American Metal, and Selection Trust agreed that each company would take one third of the venture for the purpose of bidding for the Tsumeb property. The bidding was to be done in the name of O'okiep, and Henry Smith suggested to Banghart that he write to Judge Ayer and ask for a share for O'okiep in the event that the bid were successful. This seemed only fair in view of the strong and essential support given the Tsumeb development by O'okiep personnel and its facilities. As a result, O'okiep received 9.25 per cent of the venture.

Preparing this bid was not as easy as it might seem. Many men in the three companies had strong misgivings about Tsumeb, largely because of the difficulties they foresaw in concentrating the complex and heavily oxidized ore. Regarding Tsumeb as more of a gamble than the ordinary mining proposition, the doubters were unwilling to risk more than $1 million in bidding for the Tsumeb property. Other companies had very much the same idea, with the result that all bids received by the South African Custodian were $1 million or less. In disgust, the custodian threw out all the bids and asked any companies interested to re-examine the proposition and submit new bids.

Newmont's reaction was to engage in a series of intensive and often heated arguments, discussions, and renewed investigation. Strongly supporting the Tsumeb venture were Messrs. Smith, Banghart, and Albert Livingstone, the Johannesburg attorney who had represented O'okiep since the early exploration in the 1920's. In the end, the Tsumeb enthusiasts prevailed, and O'okiep was authorized to submit a new bid of £1,010,000, or $4,040,000 at the prevailing rate of exchange.

No other bid even approached this level, and the group represented by O'okiep thereupon received all rights to the Tsumeb property. Arthur Storke, in return for initiating the idea, received 5 per cent of the venture, for which he paid $50,000, an investment that yielded millions. Storke was killed in an airplane crash in 1949, together with E. T. Stannard and Russell Parker, of Kennecott, just after Storke had been made president of Kennecott Copper Corporation.

Shortly after the transaction was completed in 1947, Banghart went back to Tsumeb to take inventory of equipment at the property. His original notebook contains a tabulation of mill equipment, consisting of three Blake crushers, hand-fed and driven by flat belts from line shafting, a tube mill, six tables, twenty-five buddles (something Banghart had never met before), two jigs, and a junkpile. Adjoining this tabulation, Banghart disgustedly scribbled: *"All* junk, value £1."

However, there was enough equipment available so that with a minimum of additional equipment, the new owners were able to build a simple sorting and gravity concentration plant, which, in March, 1947, went to work on the ore dumps. From these Tsumeb recovered concentrates, in 1947 and 1948, worth over $7 million. The flotation plant was completed in March,

1948, and the gravity plant was shut down. Actually, the Tsumeb flotation mill was built out of profits from the early operations of this inexpensive gravity concentration plant.

By mid-1948, the syndicate had put £1,207,807, plus the purchase price, into bringing the mine into full operation. Newmont's share of this sum came to £461,000, or $1,855,784. Repayment of notes reduced this investment to a net of $579,217. Banghart says that no mine he knows of ever returned so large a cash flow for such a relatively small investment as did Tsumeb.

Regarding the Tsumeb geology, Henry Smith reported that the ore minerals are associated with a breccia pipe of pseudo-aplite or pseudo-quartzite. The pipe is in general about 250 feet wide and 600 feet long. It occurs along the flank of a syncline in a thick series of pre-Cambrian dolomites. In section, the pipe dips 55 degrees south from the surface down to 2,200 feet. At that point, the pipe reverses to 75 degrees north dip, which Smith reported "keeps it conveniently within the claim boundaries." The very high-grade massive ore occurs around the periphery of the pipe, but toward the center, the ore is lower in grade, occurring in irregular veins and pods in the fractured dolomites.

Above the 1,200-foot level, the Tsumeb ore has been heavily oxidized by ground water, resulting in a complex of carbonates, sulfates, arsenates, vanadates, and oxides of copper, lead, zinc, and other metals. Between the 1,200- and 2,600-foot level, the ore consists largely of unaltered sulfides, but below the 2,600-foot level, an aerated water-bearing zone has cut into the orebodies from the north, and oxidation is widespread in the orebody's lower levels. All the present production of ore from the Tsumeb mine is regarded as oxide ore, although it does contain major amounts of sulfides as well.

Thus far, 128 different minerals have been identified in the Tsumeb orebody. Those of greatest economic importance are: chalcocite, bornite, tennantite, malachite, native copper, cuprite, galena, cerussite, sphalerite, native silver, germanite, and renierite (the last two, germanium minerals), most of which can be made to respond well to flotation.

Also important but not easily floatable are the following: conichalcite,[1] duftite,[2] mottramite,[3] mimetite,[4] anglesite, smithsonite, and willemite. It is no trick for a mining engineer to identify the last three minerals, but if he can give the composition of the first four, he is either unusually well read, or an employee of Tsumeb.

Since 1947, the Tsumeb mine has been developed down to the 38 Level, which is 4,150 feet below the surface. The shaft system is now being deepened to the 46 Level and from a drift on the 44 Level, deep drilling will be done to explore the orebody further. In the lower levels, the pipe has shown signs of narrowing.

For a production of about 50,000 tons of ore per month, the mine uses a classic cut-and-fill stoping method, and with square sets used to recover pillars, ore extraction is practically 100 per cent. In recent years, sand fill from mill tailings and quartzite sand from a surface deposit have been used to provide ground support. Because mine water is used for domestic purposes in the town of Tsumeb, a water purification plant eliminates any metal content of the mine water before it enters the domestic water supply system.

In March, 1948, one year after the first staff members had arrived at Tsumeb, the new Tsumeb mill was under cover, and the first grinding and flotation section was running. Because only medium-sized electric motors were available in South Africa, the Tsumeb mill crushers and ball mills had to be designed as small units. Therefore, the original 900-ton-per-day mill had to be designed for three circuits of 300 tons each. Instead of being a handicap, this turned out to be an essential item, when

---

[1] A basic arsenate of calcium and copper.
[2] A basic arsenate of lead and copper.
[3] A basic vanadate of lead, copper, and zinc.
[4] A chloride-phosphate-arsenate of lead.

both sulfide and oxide ores had to be treated by flotation separately.

Present capacity of the Tsumeb mill is about 2,000 tons per day, and because of the oxidized nature of the ore, the flow sheet is one of the most complicated, and the reagent schedule is one of the most extensive, anywhere in the industry. Over the years of operation, many different flow sheets were devised to cope with changes in the ore. For those interested in details, the present beneficiation practices at Tsumeb are described in a paper written by Messrs. Boyce, Venter, and Adam of the Tsumeb staff. It appears in a volume of the American Institute of Mining Engineers, including papers presented at the fall meeting of the Institute in St. Louis in September, 1970.

In addition to copper, lead, zinc, cadmium, and silver, the Tsumeb ore also carries a recoverable content of germanium. In 1952, Frank McQuiston, who had spent the previous four years working with the Atomic Energy Commission on the development of uranium recovery systems, returned to Newmont and thereafter aided the Tsumeb staff in developing a recovery procedure for Tsumeb's germanium, which had suddenly come into great demand for the manufacture of transistors.

Using a combination of flotation and magnetic separation in ferro-filters, a concentrate was produced carrying about 0.39 per cent germanium. At first, this concentrate, amounting to about 1,000 tons per year, was sent to Belgium for treatment, but research at Tsumeb and in the United States eventually developed a leach-evaporation-distillation process for producing germanium oxide from a low-grade flotation concentrate. Tsumeb built a suitable plant to carry out this process in 1960, doubling the capacity in 1962. The end product of this plant was a white, granular powder assaying 98.7 per cent germanium dioxide. To provide sulfuric acid for this plant and for the cadmium leaching plant, Tsumeb built a sulfuric acid plant in 1962.

In the early days of the operation, Tsumeb shipped only metal concentrates or direct smelting ore. At no time, however, were the Tsumeb concentrates eagerly sought after by any of the world's copper or lead smelters. Because of contained impurities, smelter charges for Tsumeb concentrates were high and penalties were severe. The copper concentrate carried 43 per cent copper, 9.7 per cent lead, 8.2 per cent zinc, and 4 per cent arsenic. The lead concentrate carried about 46.8 per cent lead, 4.8 per cent copper, 6.4 per cent zinc, and 1.8 per cent arsenic. These assays indicate that in purchasing the Tsumeb concentrates, a copper smelter would have to treat materials carrying more lead, arsenic, and zinc than any other concentrate it normally would buy. Likewise, the lead concentrate carries more copper and arsenic than any lead smelter outside Tsumeb is accustomed to handle.

Therefore, high smelter charges and increasing shipping costs made it imperative that Tsumeb consider smelting its own concentrates. A study of the problem convinced the staff that it could be done, at least in copper. Late in 1959, the Tsumeb board of directors authorized construction of a copper smelter, and clearing of a site began immediately. However, before construction began, a decision was made to build a lead smelter as well, and a new site for an integrated smelter had to be found.

For the copper smelter, the decision was taken more or less to copy the O'okiep installation so far as the reverberatory furnace, converter, and crane aisle were concerned. Arrangements were made with Dorman Long of South Africa to do the detailed drawings with specifications laid down by the Tsumeb staff. Dorman Long were also commissioned to do the total steel erection and installation of equipment for both smelters, with emphasis on getting the copper smelter started as soon as possible. This arrangement proved highly satisfactory, and the copper smelter was blown in on November 2, 1962.

In the meantime, Frank McQuiston was asked to supervise the design of the lead smelter and the auxiliary equipment common to both plants. He chose Western Knapp Engineering of San Francisco to do the detailed drafting, with emphasis on completing the design work in those sections of the plant required by the copper smelter for an early start up. McQuiston also organized a team of consultants, including R. Hermsdorf, W. P. Mee, and A. Labbe to do the fundamental design of the lead smelter. This team, based in San Francisco, in cooperation with Western Knapp Engineering and the Tsumeb staff, was able to complete the design of the Tsumeb lead smelter in an incredibly short time. Construction proceeded as soon as drawings were approved and released in any particular area, and thus when the last of the 750 drawings required for the lead smelter was finished, the construction for the project was some 75 per cent completed. The lead smelter was blown in in November, 1963, some three and a half years after the selection of the site and the firm decision to proceed with the project.

Because of the unusual problems met and overcome, copper and lead smelting at Tsumeb are well worth describing in detail, although only a summary can be given here.[5]

In the copper smelter, the first major problem was not a metallurgical but a people problem, in that the Ovambo labor went on strike shortly after the smelter started up because they were frightened of pyrometallurgical processes hitherto unknown to them. For a short period, Charlie Stott, the Tsumeb general manager, swept the scales in the copper smelter, while his assistant, James Ratledge, pushed a bullion barrel. Once the strike was settled, the major problems resolved into removal of impurities from the blister copper, the prevention of a severe build-up of magnetite in the reverberatory furnace, and the handling of slag too high in metal content to be discarded.

Consisting mainly of lead and arsenic, the impurities could be slagged off by an oxidation process, which the Tsumeb staff discovered could be done in the converter itself. However, the oxide slag thus formed is explosive if not handled carefully, and furthermore, brings great quantities of magnetite back to the reverberatory furnace, which had to be dug out after less than a year's operation. Reverberatory slags were then assaying up to 2 per cent copper and 8 per cent lead. The problems of magnetite

[5] Description of Tsumeb smelting furnished by Peter Philip, assistant general manager.

Spectacular Tsumeb outcrop as it was in 1905, consisting mainly of quartz in a ridge 40 feet high, shot through with green and blue oxidized copper minerals. *Photos courtesy South West Africa Scientific Society*

Before the railroad reached Tsumeb in 1906, high grade ore from just below the outcrop was shipped out to Swakopmund in wagons hauled by oxen.

Pouring lead bullion into a dross kettle in the Tsumeb lead refinery. Ted Hurley, lead plant superintendent, watches. Only two electrolytic lead refineries in the world produce lead exceeding Tsumeb's in purity.

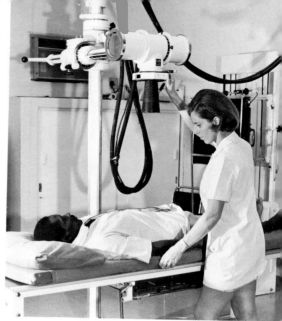

LEFT: Aptitude testing for new employees is part of Tsumeb's training center activities. Truck driver candidate gets two-hand coordination test. RIGHT: O'okiep's fully equipped hospital at Nababeep offers complete facilities for all employees. Here, Mrs. Speirs, radiographer, prepares for X-raying of patient.

Modern load-haul equipment like this machine helps O'okiep cope with rising costs. Non-white workers increase earning power by qualifying for such jobs.

O'okiep Copper Company Limited's mine, mill, and smelter as they were in 1938, when George Kervin, Wes Goss and Gene Tucker were preparing the property for production.

Design problem at O'okiep's Carolusberg mine occupies, from left, John Nangle, general superintendent (mines), Gordon Parker, O'okiep general manager, and Jack Ferguson, now manager of the Matchless Mine of Tsumeb.

build-up on the reverberatory furnace bottoms and the poor quality of the slag were both solved by sending the converter slag, which contained nearly 30 per cent lead, directly to the lead smelter, following granulation.

The Tsumeb copper reverberatory furnace produces a very high-grade matte, and to provide heat enough in the subsequent converter operation, the company had, at first, to buy about 1,000 tons a month of pyrite. Improved technique and the use of a larger blower eliminated the pyrite requirements. These were the major problems in the copper smelter, but Tsumeb's lead smelter presented a much more formidable array of difficulties.

In lead smelting, it can be taken as axiomatic that the key to successful operation of a lead blast furnace lies in the sinter plant, not in the blast furnace itself. This is especially true of a lead smelter that has just installed the relatively recent practice of updraft sintering. Because Tsumeb was only the second lead smelter in the industry to install a large-scale updraft sinter machine, the Tsumeb staff had to learn this axiom the hard way. Within twelve hours of initial start-up, the Tsumeb blast furnace froze tight. Digging out a frozen lead blast furnace is a costly and laborious experience that no one is anxious to repeat, yet the Tsumeb staff faced this ugly job again and again. It was only after much study and experimentation in sinter practice and many changes in the sinter plant that the Tsumeb staff was able to attain acceptable blast furnace performance and release from the fear of imminent furnace freeze-up.

Essentially, the quality of the sinter was improved by careful sizing of sinter plant fluxes, including rod mill grinding of all fluxes and secondary materials going to the sinter machine. In addition, the staff discovered that it was imperative that return sinter be accurately sized in order to create nuclei as sintered cores for the growth of pellets for sinter machine feed. This step created a sinter bed of optimum permeability. These, and other changes developed over a period of some months, resulted in production of sinter of a quality sufficiently good to smooth out the blast furnace operation.

Because the Tsumeb lead concentrate carries an unusually high content of copper and arsenic, the Tsumeb blast furnace produces a great deal of copper matte and speiss (a copper-lead arsenide) in addition to the lead bullion. First, the separation of speiss and matte was very poor, and the two small reverberatory furnaces provided for smelting lead drosses could not handle the load of dross due to poor combustion in the furnaces. This was rectified by installing seven small producer gas burners per furnace instead of two large burners each. This change allowed all dross to be smelted in one furnace. Careful control of the copper–arsenic ratio in the sinter plant also eliminated excessive build-up in the primary dross kettles, thus contributing materially to the increased capacity of the dross plant.

Eventually, the staff found that by placing soda ash in the bottom of each 35-ton ladle, and allowing the charge to settle briefly, speiss and matte could be caused to collect in the top one third of the ladle's charge. This upper layer was then poured directly into the dross furnace, and the remainder of the charge went to the dross kettle. This step has substantially cut consumption of soda ash, and reduced labor and crane time, and now only half the kettles in the drossing plant need be used.

Speiss and matte are now granulated together, and are roasted with pyrite from the Matchless mine for arsenic removal. Roasted dross skims now go to the copper smelter as a result of this operation, which was at one time thought to be impossible.

To summarize, the following are the major developments incorporated in the Tsumeb lead smelter:

1. Automatic fingertip-controlled proportioning of sinter feed materials.
2. Automatic proportioning of blast-furnace feed material, and furnace charging by conveyors.
3. Continuous blast furnace tapping (not a Tsumeb innovation, but obtained on royalty from American Smelting and Refining Company).

4. Improved dross handling techniques for treatment of an exceptionally high dross fall.
5. A crushing and screening plant to crush, size, and distribute a great variety of products to the sinter plant, the blast furnace, and the copper reverberatory furnace.

Considering that it began work on a uniquely unsatisfactory lead concentrate, the Tsumeb staff is understandably proud that the development work in the lead smelter has enabled Tsumeb to produce a lead bullion that contains fewer impurities than the bullion produced by nearly any other lead smelter in the world. The only exceptions to this are the Betts electrolytic lead refineries operated by the Cerro Corporation in Peru and by the Consolidated Mining and Smelting Company of Canada in British Columbia. Tsumeb lead consistently runs 99.998 per cent lead, and on occasion it has been known to reach a purity of 99.999 per cent.

Operations at Tsumeb are by no means static. In 1965, the company established a large research and development center at Tsumeb, intended largely to work on operating problems. Recently, however, research has centered on investigation of a zinc fuming process to treat lead blast furnace slag, as well as on development of a new germanium recovery process. It seems clear that from sheer necessity, the Tsumeb staff has developed a capability for handling complex concentrates by smelting or other procedures that is probably not duplicated in any other smelter in the world.

Along with its activities at Tsumeb proper, the company has conducted an intensive campaign of exploration and development elsewhere in South West Africa. The first result of this campaign was the Kombat mine, located about 65 miles by road from Tsumeb in the Otavi Valley. The Kombat mine has copper-lead orebodies once worked by the Germans, but abandoned when the mine workings encountered heavy water flows. Later, drilling done by Tsumeb disclosed ore reserves that, as of July, 1971, totaled about 1,740,000 tons of ore assaying 1.88 per cent copper and 2.70 per cent lead. Production of concentrates began in 1962, and at present, the Kombat mill treats about 1,400 tons of ore daily, producing clean concentrates of both lead and copper that are quite helpful in the Tsumeb smelter. These concentrates carry some silver, but no zinc, cadmium, or germanium.

A second result of Tsumeb's exploration effort is the Matchless mine, located almost 300 miles by road from Tsumeb and only 26 miles from Windhoek. The Matchless mine has a copper orebody that had been mined as long ago as 1900. However, once the initial miners got below the zone of secondary enrichment, they could find no more direct shipping ore, and the property was abandoned. Tsumeb's drilling campaign has disclosed ore reserves totaling about 2 million tons assaying 1.81 per cent copper, and 13.02 per cent sulfur. Chalcopyrite and pyrite are the ore minerals.

At a cost of about $5 million, Tsumeb developed the Matchless mine and built a mill with a capacity of 500 tons per day. The mill began operating in June, 1970, at 250 tons per day, and in the fall of the year, when the shaft was completed, the milling rate was increased to 500 tons per day. However, mining became unprofitable at 1971's metal prices, and in January, 1972, following the strike, the Matchless mine was shut down in order to concentrate on development. It was expected that production might resume by late 1972.

Aside from its technical accomplishments, Tsumeb is noteworthy for the attention the company has paid to providing modern community facilities for all its employees. After all, no one is going to be happy living and working in a remote, inhospitable area unless he is comfortable, and therefore the company provides not only houses and all ordinary community facilities, but complete recreational amenities as well. Tsumeb operates one African school in the nearby farm location and has built and donated a second school to the administration, located at Tsumeb. The capital investment in the Tsumeb venture adds up to $70,894,000, all provided out of cash flow of earnings, of which only about $4 million was used to repay a short-term loan, which was borrowed from a bank to supplement the funds required for smelter construction. Of this amount, $13.5 million has been spent for townsite, housing, vocational, and recreational facilities. A sense of community responsibility has been characteristic of Newmont wherever it has been involved in mining management, and in few places has this attitude been more evident than at Tsumeb.

Early in the venture, the Tsumeb management realized that a major problem was going to be a lack of manpower, skilled or unskilled. Most of the local population lived in homelands where conditions were much the same as they had been for centuries. In their homes, the Ovambo were living under conditions of abject poverty. Disease was rampant, education was almost nonexistent. The Tsumeb Corporation at once set about providing improved living conditions for its employees from Owambo. Today there are about 4,000 Africans working at Tsumeb and Kombat, most of them receiving on-the-job training. In addition, the company has recently provided an aptitude testing and training center to assist the Ovambo to develop greater skills and earning power.

The workers live in clean, cool, stucco quarters. A large beef herd is maintained to provide fresh supplies of meat. A modern kitchen prepares meals for the workers, and the 4,500-calorie daily diet for each man results in an average weight gain of from twenty to thirty pounds during a man's first year on the job. Tsumeb has a modern hospital staffed by qualified doctors and nurses and carries on a preventive medical program as well.

All in all, Tsumeb represents an accomplishment of which everyone involved can and does feel proud. Since 1947, Tsumeb has had three managing directors, Henry DeWitt Smith until 1953, Marcus D. Banghart until 1969, and currently, David O. Pearce. The first men on the job were Dave Christie, metallurgist, and Bob Cockburn, mine secretary, who were transferred from O'okiep to Tsumeb in January, 1947. Christie supervised the sorting and shipment of the first carloads of ore, which ran over 45 per cent lead. In March, 1947, Christie and Cockburn were joined by Jack Ward, manager, Frank Pickard, mine superintendent, Jack Ong, mill superintendent, and Howard Boyce, mill foreman. These were the men who got Tsumeb into production so effectively and so rapidly that the company paid off its debts and declared its first dividend by June 1, 1950. Today, Jim Ratledge is general manager at Tsumeb, Dave Christie is in the Newmont office in New York, and Bob Cockburn is secretary for Granduc Operating Company in British Columbia. Boyce retired as general superintendent of Tsumeb in early 1971.

An accomplished mineral dressing engineer and metallurgist, Jim Ratledge was born in Canada and acquired his first mill operating experience at the Bralorne mine in British Columbia. Giving up his original vocation of school teacher, he went to South Africa in 1936 to work in one of the large gold mills near Johannesburg. In 1938, he went to the Macalder gold-copper mine in Kenya, using his earnings from this job to put himself through Queens University for a degree in metallurgy in 1946. Shortly thereafter, he very nearly went to Northern Rhodesia, but an acquaintanceship with Henry Smith and Marcus Banghart led to his joining the original Tsumeb staff in 1947.

Ratledge now heads an enthusiastic staff, all of whom take an obvious pleasure in the job they are doing at Tsumeb, despite the fearsome technological problems with which they must continuously contend. In fact, Tsumeb seems to arouse an affection and enthusiasm in men not otherwise given to emotional outbursts. For example, one of the Tsumeb officers has this to say about the South African operations: "If it had not been for O'okiep and Tsumeb, Newmont would still be a small company, operating a lot of little gold mines here and there." A strong statement, perhaps, but an equally strong indication of loyalty and pride, which, like all oversimplifications, ignores certain facts, such as the contribution of Magma, Idarado, and Newmont's portfolio of mining and petroleum blue chips.

What, then, of the future? It is true that at Tsumeb ore reserves are in a slow decline, inasmuch as production recently has exceeded development of new reserves. As of July, 1971, Tsumeb had ore reserves of 5,950,000 tons running 4.52 per cent copper, 9.37 per cent lead, and 2.40 per cent zinc. In addition to the work at the Kombat and the Matchless properties, exploration elsewhere in South West Africa is being pursued actively, but without significant results as yet.

Tsumeb, for the year ending December 31, 1971, sold metals for a value of $47,735,000, out of which it was able to declare a dividend for the year of $14 million, or $3.50 per share on the 4 million shares outstanding. Newmont's direct interest in Tsumeb is still 29.2 per cent.

Tsumeb earnings, plus substantial income from O'okiep and elsewhere, make it clear that Newmont made no mistake many years ago when the company decided to emphasize strongly mine-seeking outside the United States. Actually, these foreign dividends made it possible for Newmont to consider and to participate in or carry out such major projects abroad as Southern Peru, Palabora, and Highveld, and in North America, Magma, Granduc, Similkameen, Atlantic Cement, and others. To put it another way, by going abroad, Newmont acquired the financial strength to expand and, lately, to emphasize its holdings in North America in a whole series of major projects that have begun to reach fruition in only the last two or three years. Succeeding chapters will show how this was done.

The Silver Queen Mine, 1910

8 Magma — Growth at Superior

Having succeeded, with O'okiep and Tsumeb, beyond all expectations, Newmont might reasonably have been expected to devote even more of its efforts to foreign development. After all, with the African cash flows as a base, Newmont could get out its high-caliber rifle and become a hunter of big game. And where, one might ask, were bigger game to be found than in Africa?

With the clear vision of hindsight, we could now answer "Arizona," but in those confused, turbulent postwar years, that answer was apparent to only one or two men within Newmont and Magma. Not until the late 1950's or early 1960's did a degree of concern begin to develop in tne company over what some members of the board and the management thought might be an overemphasis on foreign mineral development.

Therefore, it is probably quite fortuitous that in the decade between 1947 and 1957, Newmont launched major involvements, or made substantial investments, in ten enterprises in the

United States and Canada, but made only six such investments in foreign enterprises.

The ten North American investments included Sherritt Gordon, Western Nickel, Granduc, Cassiar Asbestos, and Dawn, together with investments in five natural gas and gas pipeline companies. The gas company investments, which proved quite profitable, were undertaken in the belief that gas and gas transmission companies would be less affected by rapidly rising labor costs than would more labor-intensive projects.

The foreign investments in that period include Tsumeb, Palabora, Atlas Consolidated, and Southern Peru, all of them major enterprises, but in the long run they became outweighed in relative importance for Newmont by the North American ventures. Hardly visible as a trend at first, the shift away from strong emphasis on foreign operations became in the 1960's a firm company policy.

Oddly enough, the most effective instrument for accomplishing this shift was the company that Colonel Thompson incorporated in 1910, Magma Copper Company. Beginning as a worked-out silver mine bought by Colonel Thompson for $130,000, Magma, sixty-two years later, represents a total investment of funds, partly borrowed and partly self-generated, approaching $450 million. Since production of copper began in 1915, Magma and its San Manuel division have produced approximately $1,480,642,000 in copper, zinc, molybdenum, gold, and silver through 1971.

Although such growth must obviously have been the work of many men, Magma owes more of its achievement to Alexander J. McNab than to any other individual. Almost alone among his associates in Magma and Newmont in this respect, McNab had never really lost his faith in the United States and Canada as the real source of Newmont's strength.

However, as pointed out previously, despite what appeared to be a close relationship between the two companies, only in very recent years could Magma be considered a "Newmont" operation. Magma was an operating, publicly owned company long before Newmont was incorporated, and ownership might well have passed into other hands, as did Inspiration, Kennecott, and other Thompson promotions, except, one supposes, that the colonel had a special affection for Magma and the Superior area, where he established an arboretum and built his western home, Picket Post House.

It certainly was not that Thompson looked on Magma as a great money-maker. His own holdings of Magma fluctuated with the stock market price and were never very large. It is said that one of his last bits of advice to his family, which wasn't heeded, was to sell Magma, because he thought that the mine was about worked out. McNab never thought so, and McNab was right. In following the story of Magma, one should keep in mind that not until the 1960's was there any significant merging of the pathways followed independently until then by Magma and Newmont.

Symptomatic of this independence of Magma was the fact that Fred Searls, from 1925, when he joined Newmont, until very late in his career, never became closely identified with Magma. Searls did eventually regard himself as the nominal chief geologist of Magma, and he would go out to Arizona now and then to poke around in the mine. However, his real interests lay elsewhere, and Magma was not one of his enthusiasms.

Magma's beginning was most unimpressive. A tottering wooden headframe, some ramshackle buildings, and a skimpy ore dump were all that marked the Silver Queen mine when Thompson bought it on Henry Krumb's rather tentative recommendation back in 1910. Krumb thought that it might be of interest for its copper content rather than the remaining indications of silver. The Silver Queen, now metamorphosed into the Magma mine, was located at the foot of a forbidding rocky wall known locally as the "Apache Leap," an escarpment, apparently blocking the way of a traveler bound east for Miami and Globe. Hidden from casual view, a rocky canyon opens out to provide passage. At the mouth of the canyon, the sunbaked town of Superior grew up below the Magma mine, mill, and smelter. As anyone who has spent a non-airconditioned night in Superior during the summer months knows, heat builds up in the rocky wall above the town during the day and is radiated out over the town at night. Superior is no garden spot, but it somehow seems to inspire affection in many of the people who have spent years in working and living there.

When Magma entered production in 1915, Walter Aldridge, Thompson's chief engineer, became president, H.F.J. Knobloch, vice president, and Henry Dodge, treasurer. A. J. McNab joined the company in New York in 1919. In 1915 its first year, Magma produced 44,000 tons of ore running 8.19 per cent copper and 9.57 ounces of silver. A shaft was put down to the 1,200-foot level, and the company built a flotation mill with a capacity of 150 tons of ore per day. Flotation was quite a new thing in the United States at the time, and the Minerals Separation Company of England believed it had established a world monopoly on the flotation technique. Magma was among the many who were sued by Minerals Separation for infringement, but the suit against Magma dragged on for ten or fifteen years and was finally abandoned without a conclusive decision having been reached.

Magma increased production through the years of the First World War, developing profits of over $1 million annually. A drop in the price of copper forced a shutdown of the mine in March, 1921, and plans for a smelter for Magma occupied McNab during 1921 and 1922. The mine resumed production early in 1923. Aldridge and Knobloch went off to Texas Gulf Sulphur in 1918, and later were joined by Wilber Judson. Charles Ayer became president in Aldridge's place, and McNab became vice president and a director in 1923.

Magma's smelter was completed late in 1923, and it began operating in March, 1924. Capacity of the smelter was 1,500 tons a month of blister copper. William Koerner became general manager of Magma that year, doing an extremely competent job through the 1920's.

Production continued more or less uneventfully at Magma

with a rather consistent profitability. There was nothing really easy about the mining operation, however, in that Magma's copper occurs in rather narrow veins in two main systems. The Magma vein occurs in a fault fissure possessing distinct walls and showing great persistence, both laterally and in depth. Running generally east and west, the Magma vein dips about 70 degrees to the south, and cuts through a 6,000-foot thickness of limestone, quartzite, shale, diabase, and schist in that order. These rocks strike at right angles to the Magma vein and dip 30 degrees eastward. They are covered to the east by a great thickness of volcanic rocks of comparatively recent age, which forms the rock wall blocking Superior's view to the east. Mining on the Magma vein has gone as deep as 4,900 feet.

A smaller vein, called the Koerner, has also been mined extensively. For many years, Magma's entire production came from these two vein systems. However, had it not been for discovery of a different type of ore about twenty years ago, the Magma mine might long since have been worked out.

The narrow veins and the relatively weak walls required use of square-set stoping as the mining method, followed by waste filling. Such a method does not lend itself to cheap, large-scale operations, and the high operating costs were offset only by the relatively high grade of Magma ore and the credits provided by its gold and silver.

Through 1929 and 1930, Magma appears to have suffered hardly at all from the onset of the Depression. In 1931, however, the copper price averaged $0.089 a pound and Magma's profits declined. The smelter did not operate between June and October, 1931. In 1932, the price of copper averaged only $0.0625 a pound, and with an operating cost of $0.0833 a pound, Magma lost $500,000 in the year. The mine, mill, and smelter were not in production from July to December, 1932.

Things improved somewhat in 1933, although the copper price had gone up only a fraction. The smelter was again shut down from July to December, but Magma managed to show a profit of $5,700 for the year.

A most extraordinary accomplishment of the Magma staff in

those years was a reduction of cost to what seems to us now to be an unbelievably low level. For 1934, operating cost was $0.0573 per pound of copper, which enabled the company to show a profit for 1934 of $647,000 with the price of copper averaging $0.0785 a pound. Actually, Magma managed to keep its operating costs under $0.08 a pound until 1941, when costs began to climb, reaching $0.1028 a pound in 1942.

Early in the life of the Magma mine, a problem had developed that did not become acute, however, until 1937. This was the fact that the name, Magma, turned out to be singularly appropriate for this particular mine. As the mine deepened, the rock temperature increased steadily, just as if the workings were approaching the everlasting heat of a deeply buried magmatic intrusion. On the 4,000-foot level of the mine, the rock temperature was 140 degrees Fahrenheit, a heat that rendered it impossible for men to work for more than an hour at a time. The ventilation system of the Magma has always been given extraordinary attention because of this fact of increasing temperature with depth.

However, no ventilation system could handle temperature increases of this kind. Therefore, the fairly recently developed science of air conditioning was called into play. In 1937, the Carrier Company supplied two 142-ton refrigeration units, which went into operation on the 3,600-foot level of the mine. This resulted in a drop in temperature of at least 10 degrees, which made continued operation of the mine possible. Two years later, a third unit of the same size was installed on the 4,000-foot level.

Magma was certainly one of the first underground mines to install air conditioning, although one of the mines of Anaconda in Butte is believed to have been the first. As the years have gone by and the mine has pushed downward, the air conditioning system has had to be steadily increased, and millions of dollars have been spent for this equipment. For example, in the deepest levels of the mine, even a single drift crew requires its own refrigeration unit in order to carry on work.

In 1940, William Koerner, who had been general manager

at Magma for eighteen years, died after a long illness, leaving behind him a remarkable record of achievement. He was succeeded by Edward G. Dentzer, who had been Koerner's assistant. Dentzer had hardly become settled as general manager, when the entry of the United States into the war called for an all-out production effort at Magma. At almost the same time, manpower for mining became a problem, because in the beginning, a good many men were drafted before the local draft boards began to realize that there were occupations quite as important to the war effort as that of soldiering. As the manpower problem increased in severity, production became steadily more difficult and more expensive.

In 1944, Dentzer resigned because of ill health, and Wesley P. Goss, who had just done a thorough report on the Magma operation, was brought over from the Gray Eagle mine to become general manager at Magma. Recognizing that only adequate manpower could handle Magma's particular production problem, Goss devoted most of his efforts to rounding up qualified men. At the time, the U.S. Employment Service was supposed to be the only source of labor. Goss did his best through that agency, but also tried to get men anywhere he could.

A Mexican-American society in Phoenix helped Goss get about thirty-five "wetbacks," Mexicans who had entered the U.S. illegally, and good workers they were until the U.S. Immigration and Naturalization Service got wind of it and sent them all home. Men came from the Apache reservation and from the army, by special release. Magma obtained one such group of thirty men in California, gave them bus fare to Superior, and started them off. Of the thirty men who were given tickets, probably half of them turned in the tickets for cash and disappeared. Many of the others were lost en route to Arizona, and Goss reports that only three men ever got as far as Superior. Two of them took one look at the town, turned around, and got back on the bus. The labor situation really reached its worst in 1945, but somehow Magma struggled through, and thereafter things began to improve.

Unfortunately, as an increasing number of men became

available after the war, costs also began to rise and in 1946 were double what they had been in 1939. McNab recognized that one area of improvement lay in Magma's antiquated mill, which had only been modified and not basically improved since 1915. Therefore, he had Wes Goss put in a new 1,500-ton-per-day mill, replacing the old 1,000-ton plant, with a most salutary effect on costs. It was most timely because metal prices rose a couple of years later, and the extra earnings enabled Magma to contribute around $25 million to the cost of the San Manuel project, which is a separate story, covered in the next chapter.

A most significant development at Superior came during 1947 and 1948, as a follow-up to some studies initiated by John Gustafson, a Newmont consulting geologist. Exploration to the east of the main center of mining resulted in the discovery of ore concentrations of quite a different variety than Magma had ever mined. Drifting far to the east on the 3,000-foot level of the mine, as Gustafson had suggested, the crews ran into zones of replacement of limestone by a copper mineralization, which greatly resembled the richly productive orebodies mined years previously by other companies near Globe to the east of Superior. Most prominent of these orebodies was that worked by the Old Dominion mine.

Because these replacement orebodies offered Magma the hope of larger ore reserves and therefore of increasing production and reducing mining costs, exploration in the far-east area was intensified. Over the next few years, as exploration developed additional ore reserves of the replacement type, more and more of Magma's production came from this far-east area. By 1962, approximately 90 per cent of the Magma mine's production of copper was coming from these replacement orebodies.

Initially, this replacement ore was mined by square-setting, but the method soon proved even more costly and inefficient than it was in the Magma mine's narrow veins. It was quite difficult and expensive to get heavy timbers down through small shafts and out in small cars through haulageways 5,000 to 10,000 feet long. Naturally, the staff tried other mining methods.

At first, relatively narrow limestone beds containing the replacement ore were mined by advancing at the top of the bed up the dip of about 30 degrees to the east. The roof of this opening, or as miners call it, the back, was supported by roof bolts rather than by timber. The broken ore was moved back down to the entry by means of scrapers, commonly called slushers. However, in the thicker parts of these orebodies and in areas of highly fractured backs, higher-cost square-set stoping still had to be used, and costs of production went sky high in the 1960's, in keeping with the rapidly rising wage scale. Consequently, the staff devoted a great deal of time to experimentation to develop a more suitable mining method. Nevertheless, it became increasingly evident that a low level of production, combined with high indirect costs occasioned by small and distant entries, antiquated equipment, and excessive burden of air conditioning and ventilation would militate against any successful attempt to reduce costs.

Fortunately, in 1965, diamond drilling from replacement zones already mined out disclosed even thicker and larger replacement orebodies in limestone some 400 feet stratigraphically above the areas previously mined, as well as in two other limestone strata above the original replacement orebody. Further exploration over the next two years developed a proven reserve of about 10 million tons of ore running 5.8 per cent copper. For the first time in the life of the Magma mine, sufficient ore reserves were indicated to justify large capital expenditures, initially estimated at $31 million, for enlargement of productive facilities. Consequently, the management decided to increase mill capacity from 1,500 to 3,000 tons of ore per day and to provide a more efficient means of transporting the ore out of the mine and up to the concentrator.

In 1971, the scope of the Magma mine project was enlarged to include a mill with a capacity of 3,500 tons daily, and it was decided to move the mine's entire surface plant and facilities from the old site at Superior to a new shaft site at the top of the escarpment. These and other changes, plus spiraling construction costs, pushed the total cost of the Superior expansion project up to an estimated $75 million. Completion is expected late in 1973.

A new, larger-capacity shaft was required for the expansion, both to handle the increased tonnage and to provide improved access to the eastern orebodies. The new shaft will be 4,100 feet deep and will combine with a 9,000-foot tunnel, 12½ feet in diameter, to cut transport time and costs between surface and working faces. Travel time for miners in the new system will be reduced by an average of about 2.5 hours per shift.

Further to increase productivity and to reduce costs, the staff developed a new mining method that would work efficiently in the ore zones in the new area, which are much thicker than those first discovered years before.

As it was developed by Magma's staff mining engineers, the new method is basically an adaptation of a system used in Canadian nickel mines. The method is usually referred to as a modified undercut-and-fill technique. Used in limestone ore beds of about 100 feet in thickness, the method involves driving across the top of the orebody, extracting the ore through ore passes, and then filling the drive by classified mill tailings containing cement in a ratio of 1 part cement to 20 parts sand. When this mixture has set up, a second drive proceeds immediately below the filled drift under the sand-cement back.

Proceeding in a series of panels across the ore zone, leaving pillars between, and then going back to extract the pillars in the same way, the miners gain practically 100 per cent extraction of the ore. Furthermore, the output per man-shift is appreciably higher than that of a square-set stoping method. Timber consumption is cut at least in half. It is not too much to say that this new method, in combination with the method developed for the narrower beds, will greatly prolong the life of the Magma mine.

To provide a more efficient means of transporting ore from the mine to the surface, air operated, pneumatic-tired loaders are used in the mining drifts to move ore to the ore passes. From the ore passes, ore goes to the new shaft to be hoisted to the new tunnel. Incidentally, this tunnel was driven by a tunnel-boring machine. In limestone, this machine could drive 50 to 60 feet a day of tunnel 12.5 feet in diameter, although in quartzite the advance was slowed materially because of the hard and extremely abrasive nature of the rock.

When the expansion of San Manuel's smelter was completed, the Magma smelter at Superior was shut down, and after August, 1971, concentrates from the enlarged Magma mill were shipped down to San Manuel for smelting.

In even this brief survey of the history of the Magma mine of Magma Copper Company, there have been evident the three outstanding skills of Magma and Newmont, namely exploration, technical, and financial competence. In addition, Magma's management exhibited stubborn perseverance in finding additional ore reserves and in the faith that kept McNab and his associates driving ahead in the face of extreme adversity. Even though the Magma mine at Superior has in recent years been overshadowed by the San Manuel development, the men at Superior can look back with pride on a record of outstanding accomplishment.

The San Manuel Mine, 1955

9 Magma—Determination
Rewarded at San Manuel

Magma Copper Company came into existence at roughly the time when metal mining in this country had just begun to change in character from a kind of lively gamble to the respectable, fairly predictable, not nearly so exciting activity it is today.

Since the time when Henry Krumb began developing accurate methods of sampling for evaluating large porphyry copper orebodies, mining companies have developed a whole science of evaluation and prediction, drawing on the skills of geophysics, geology, engineering, economics, and finance to reduce to a minimum the risks involved in a large, new mining venture.

Does this mean that risks in mining have now been eliminated? Are all mining ventures now so carefully analyzed and planned as to preclude the possibility of loss? Obviously not. One has to look no further than this history to find proof of that. Atlantic Cement, planning for which began in 1961, ran steadily at a loss until 1970 and 1971. Few new ventures have been as thoroughly analyzed in advance as Granduc, yet Granduc Operating Company ran at a loss throughout 1971, save for one month. Clearly, there can be imponderables that defy prediction and definition and may impair the accuracy of estimates.

However, the San Manuel project of Magma Copper Company has developed as successfully as could have been hoped, although only by extraordinary combination of technical skill, financial acumen, sound judgment, and dogged persistence on the part of Magma's management. At several critical points, a wrong decision, or even irresolution, could have ruined, or lost, the whole enterprise, but at each such point, the decisions were right and the will was firm. This chapter traces the San Manuel development from its inception in 1944 on through the difficult years of financing and construction to its solid establishment as a profitable enterprise in the 1960's. The subsequent close association with Newmont is described in Chapter 13.

In 1944, Magma Copper Company was desperately in need of additional sources of water for its concentrator. Because development at Superior had been sacrificed to some extent in the interest of all-out production for the war effort, A. J. McNab asked Newmont if he could borrow John K. Gustafson, a Newmont geologist, to aid in planning a resumption of exploration and development work that hopefully would find additional reserves of both ore and water.

Gustafson arrived at Superior in April of 1944. Along in June, he was going over some maps in the Magma engineering office when he overheard a snatch of conversation outside the door.

"I hear the Bureau of Mines is going to drill Nick's copper prospect" said someone.

"That's right" was the reply. "I understand they've put down one hole that had one per cent copper in it."

Intrigued by this comment, Gustafson inquired in the engineering office, "Who is Nick and where is his copper prospect?" He was told that the "Nick" referred to was Henry Nichols, an assayer for Magma, and that the prospect was the old San Manuel property down near Oracle. He was also shown a report on the claims that was written in 1941 by a Magma engineer, there being no geologist available at the time.

According to the report, the San Manuel claims were without value for Magma, and at this point, Gustafson made one of those make-or-break decisions that can radically alter a company's whole future. Gustafson could have said to himself, "Well, it really doesn't look so hot. If the Bureau does find something, we will probably hear about it," and then have gone back to the job he was doing at the time. Instead, Gustafson had a hunch there might be something worthwhile at San Manuel, and he went looking for Nichols.

He found Nichols reluctant even to discuss San Manuel. After having offered the property to Magma more than once and having been turned down flatly, Nichols had the attitude that Magma was "dead on its feet," and he wasn't about to expose himself to another brush-off. Gustafson persisted, however, and eventually Nichols thawed out enough to take Gustafson on a tour of the claims the following Sunday.

When Gustafson got back to Superior that night, he told his wife that he had seen "one of the best raw prospects I ever encountered." The reason for this optimism was that Gustafson knew that most big porphyry copper deposits had relatively barren rims of pyrite mineralization. Red Hill, the prominent San Manuel landmark, apparently could be one of these big rims. He noted that the Red Hill was a leached outcrop of pyrite mineralization and that the only limonite derived from copper mineralization occurred at the base of the hill just before the rocks were covered for miles by the Gila conglomerate. Sticking up through the Gila conglomerate were two desk-sized outcrops of oxidized copper that looked as if they would assay about 1 per cent copper. Set up on one of these outcrops was a Bureau of Mines drill, which was then down about 40 feet.

The next day Gustafson telephoned McNab in New York

and got his O.K. to try to obtain the property for Magma. He told Ed Dentzer, the general manager at Superior for Magma, that he would try to get an option on the property.

Gustafson then started negotiating with Nichols and his three partners, Messrs. Douglas, Giffin, and Erickson, for an option on the property. The situation was ticklish and complicated in that James Douglas and Erickson had married sisters. Douglas's marriage had ended in divorce, and there was bad blood between the two. There was also a question of percentage ownership to be resolved. Finally, the partners were reconciled sufficiently for Gustafson to get a verbal agreement to an option on the property, which was later modified by renegotiation after Wesley Goss appeared on the scene. Wesley Goss replaced Ed Dentzer as general manager in July, 1944. After his arrival, Gustafson and Goss worked together planning the rest of the acquisitions and ultimate exploration of the property.

In August, 1944, Gustafson and Wesley Goss worked out a revised agreement for the property acceptable to all the partners. Just before the agreement was signed, another mining company was said to have offered Douglas $100,000 cash to "forget" his verbal agreement with Magma and sign up with them. Douglas said he had too good a memory and declined.

In his report on San Manuel, Gustafson mentioned that beginning early in 1944, the U.S. Bureau of Mines had drilled the San Manuel prospect at the suggestion of the U.S. Geological Survey, which, at the request of the War Production Board, had been attempting to develop additional domestic sources of copper. By the time Gustafson heard of the project, the Bureau had put down six holes, one of them as deep as 1,000 feet, which had encountered 1 per cent copper in the primary sulfide zone.

The first San Manuel claims were staked back about 1870 over a small area surrounding Red Hill. Wes Goss learned later that one of Colonel Thompson's crews actually did some drilling at San Manuel in 1917, but had become discouraged because of the barren pyrite mineralization. No mining activity has ever been recorded at the property prior to opening the San Manuel mine.

Prior to Gustafson's learning of San Manuel, no one in the New York offices of either Magma or Newmont had ever heard of the property. The results of the 1917 drilling had been filed and forgotten at Superior. Neither Fred Searls nor Henry Smith, who were away in Washington at the time, had been consulted about the Bureau of Mines drilling campaign at San Manuel, nor had either one any indication that such a program had been started.

In 1925, Anselmo Laguna, of Superior, began to acquire the claims around Red Hill and over the years managed to stake most of what eventually turned out to be the best ground. He also acquired a partner, James Douglas, owner of a saloon in Superior. Although they could never interest anyone else in their claim, the partners did develop a real friendship. When Laguna, practically penniless, became seriously ill in the late 1930's, Douglas cared for him and paid his medical bills. As one of his last acts, Laguna deeded all his Red Hill claims to Douglas.

Later, Douglas sold a part interest to Burns Giffin, owner of a garage in Superior, and to Victor Erickson, a mechanic who had once worked for Magma. Giffin had a truck, useful for getting out to San Manuel, and Erickson agreed to do the assessment work on the claims.

At one time the partners offered the claims to Magma for $50,000. Magma sent down an engineer who reported that San Manuel had no apparent value.

In 1940, the partners approached Henry W. Nichols, a Magma assayer, with an offer of partnership in exchange for analyzing samples taken at San Manuel. Nichols thought well of the property and tried several times to interest Magma in taking up San Manuel. One such effort resulted in the negative report on San Manuel written in 1941.

Having failed to arouse Magma, Nichols and his partners approached several other mining companies, most of them larger and more active in exploration than Magma was at the time. Frustrated everywhere, the partners applied to the Reconstruction Finance Corporation for a loan in October, 1942, and as a result, the R.F.C. recommended consideration by the U.S. Geological Survey. The drilling program just mentioned was then undertaken by the bureau at the request of the U.S.G.S. Also B. S. Butler, of the University of Arizona, actively supported the R.F.C.'s recommendation.

Having tied up the San Manuel claims, it was necessary for Magma to protect the northern, eastern, and southern extensions of the possible orebodies. Extensions to the north were covered by adjoining claims owned by Sam Houghton, a mining engineer in the nearby town of Tiger. Additional claims were owned by the Quarelli family, owners of a saloon in Winkleman, and they had optioned their claims to Houghton. The eastern and southern extensions of the deposit were staked with a great number of mining claims by Magma.

Negotiating with Houghton was difficult, as by this time many other mining companies had heard about the Bureau's work and were trying to get the properties.

Late in September, 1944, Gustafson and Houghton reached an agreement at about 2:00 A.M. in a bar in Oracle. The agreement gave Magma the best of Houghton's claims plus the option on the Quarelli claims. Gustafson drove back to Superior that same morning and before going to bed wrote up the agreement that he made with Houghton and put it in the hands of a lawyer for drafting.

Three days later he came back with the agreement on paper ready for signature. Houghton greeted him by handing him a letter from the chief geologist of International Smelting and Refining, an Anaconda subsidiary, saying that his company agreed to conclude the purchase of Houghton's property on terms that had been proposed previously.

To Gustafson's great relief, Houghton stuck by his verbal agreement and signed the papers Gustafson had prepared. Thereupon, Houghton, the Quarelli family, Gustafson, and Gustafson's twelve-year-old son, Lew, who accompanied him on this trip, went off to the Quarelli saloon for a delicious Italian dinner to celebrate the occasion. A few days later the Anaconda representative arrived in Winkleman and was successful in acquiring the remainder of Houghton's property, which Gustafson had tried to obtain also from Houghton, but had never succeeded in pinning down.

With ground secure, Gustafson suggested to the Bureau of Mines that Magma was now in position to drill San Manuel and it was no longer necessary for the Bureau to expend its funds for that purpose. The Bureau, however, decided to use its own appropriation while Magma waited on the sidelines. The Bureau of Mines expended about a million dollars drilling the property and proving up a few tens of millions of tons of ore averaging around 0.8 per cent copper.

Although the amount of ore covered by the remaining Houghton claims was not large enough to represent a separate mining enterprise, it did offer Anaconda an asset of considerable "nuisance" value. Negotiations between Magma and Anaconda regarding this additional land dragged on for several years, and it was not until 1963 that Plato Malozemoff, Roy Bonebrake, and Wesley Goss for Magma, and Vin Perry and Charles Brinckerhoff for Anaconda worked out an agreement by which Magma acquired a lease and option on the property providing for a series of rental or royalty payments to Anaconda that will not be completed until 1988. The story of these negotiations is too detailed to present here, but it does serve as an excellent example of how agreement can be reached amicably between two mining companies in an intricate and potentially controversial matter.

John Gustafson in late 1947 was asked to become the first director of raw materials of the U.S. Atomic Energy Commission, which had taken over the Manhattan Project. Fred Searls granted him a two-year leave of absence. After two years with the A.E.C., Gustafson left the agency, but instead of returning to Newmont, he accepted a position as chief geologist with the M. A. Hanna Company. After several years as vice president of

The Hanna Mining Company, Gustafson in 1961 joined Homestake as president and one year later became its chief executive officer. In 1970 he became Homestake's chairman of the board.

For advice on drilling the San Manuel orebody, McNab turned to his old associate, Henry Krumb. McNab briefly described the orebody and asked for a description of the drilling and sampling technique Henry Krumb had worked out years previously, and which he had used at Ray, Inspiration, Utah Copper, Chino, and Nevada Consolidated. Henry Krumb's handwritten description, dated September 8, 1944, is still in the Magma file.

Using Krumb's technique, under the general direction of Wesley Goss and John Gustafson, drilling began at San Manuel. Arno Winther and Grant Rubley of Miami Copper Company helped establish the technique for drilling, sampling, and assaying. When Gustafson completed his Magma report in mid-1945 he left Superior, and Hugh Steele, then chief geologist for Magma, planned and supervised the remainder of the drilling. Drilling continued as rapidly as possible until February, 1948, when it was discontinued, not because the limit of the orebody had been delineated, but because enough ore had been proven by that time to remove all doubt of the existence of a workable mining property. The indicated reserves included over 123 million tons of oxidized ore assaying 0.767 per cent copper, and 339 million tons of sulfide ore assaying 0.788 per cent copper, for a total of 462 million tons assaying 0.782 per cent copper, the bulk of it lying over 1,500 feet below ground surface.

To handle the development at San Manuel, the San Manuel Copper Corporation was organized in August, 1945. As a part of the deal with the four partners who had originally owned San Manuel, 75,000 shares of the San Manuel corporation were transferred to the partners in agreed-upon proportions. A mechanism was provided in the deal by which these shares could be exchanged for Magma shares if the partners so desired.

In 1949, Charles Ayer retired as president of Magma and A. J. McNab took his place. Henry Dodge at that time became vice president and treasurer, and Roy Bonebrake became secretary and general counsel of Magma. A year previously, Wesley Goss had been made a vice president, and in 1949, he became a director. These four men then directed Magma through the next few difficult years of establishing San Manuel as a going concern.

San Manuel did present formidable problems despite the apparent abundance of ore reserves. For one thing, the grade of ore at that time was considered to be quite low for an underground mine, and largely because of the low grade, financing San Manuel proved to be quite difficult. Estimates indicated the cost of developing and equipping the property to be in the neighborhood of $100 million, and no bank or insurance company at that time was willing to lend money for such a large mining project without adequate guarantees. In banking circles, mining is generally considered too risky to permit large loans secured only by the mining project itself. Today, however, there are several mining projects that have loans considerably in excess of $100 million, although in situations unlike Magma's at the time.

Furthermore, Fred Searls, then president of Newmont, took a very dim view of the plan for San Manuel. In his opinion, if the property were to be placed in production at all, it should be on the more modest level of no more than 10,000 tons of ore a day, which would require less capital. McNab, on the other hand, was thinking in terms of a production of no less than 25,000 tons per day, and preferably 30,000, recognizing that sufficiently low production costs for the low-grade orebody could only be obtained with large-scale operations. Searls was willing to have Newmont help with the financing, but only on the basis of the lesser tonnage. McNab was unwilling to settle for less than the full tonnage that by Magma's estimate would make the property competitive.

All during the drilling campaign and through the years of attempting to finance San Manuel, the argument went on between these two strong and determined men. Such deep disagreement could have split the office right down the middle, but although tension ran high on the fifteenth floor at 14 Wall Street, business other than the San Manuel project went on with reasonable harmony.

Nevertheless, long discussions and some dissension over San Manuel began to hamper the meetings of both the Magma and the Newmont boards. Therefore, Phil Kraft resigned from the Magma board in December, 1948. He was replaced in March, 1949, by Mrs. Peggy Hohenlohe, Colonel Thompson's grand-daughter, who later became Mrs. Morton Downey. To complete the split, A. J. McNab and Henry Dodge resigned from the Newmont board in 1948, and Dodge also resigned as vice president and secretary of Newmont. W. T. Smith re-placed him as secretary and as a director of Newmont, and Phil Kraft, Gus Mrkvicka, and Mrs. Margaret Biddle, Thompson's daughter, also became Newmont board members.

In the light of today's concern with enhancing the role of women in corporate management, it is interesting to note that Newmont has experienced an unusually strong feminine influ-ence on its board of directors. Mrs. William Boyce Thompson was a board member from 1934 until her death in 1950. Her daughter, Margaret Biddle, was a most active board member from 1929 to 1934, and again from 1948 until her untimely death in 1956. Her daughter, Mrs. Morton Downey, served as a Newmont director from 1950 until she, too, died suddenly in 1964. Mrs. Downey's daughter, Mrs. Catherine Cook, became a director in 1964, but resigned in 1970 in favor of her brother, Christian Hohenlohe. All of them took an active interest in Newmont's affairs, and the board was the better for their pres-ence and their influence.

To provide funds needed for the drilling and initial under-ground exploratory mine development at San Manuel, Magma arranged a loan in July, 1947, of $3.5 million, shared equally be-tween Bankers Trust Company and The Mutual Life Insurance Company of New York. In August, 1952, the loan was entirely repaid.

Magma also discussed financing the entire San Manuel project with Sloan Colt, the chairman of Bankers Trust, and Stuart Silloway, who was then head of the loan department of Mutual Life. Silloway remained unconvinced, and turned down the suggestion of a really big loan. He would have been more receptive to the idea if Metropolitan Life had been willing to join in the proposition, but the banks and insurance companies generally thought San Manuel much too risky. The orebody was too deep and the ore grade too low to fit what was then the conventional pattern of successful porphyry copper mining projects.

Incidentally, Stuart Silloway later became president of Investors Diversified Services (now a director), and has been a Newmont director since 1959.

In February, 1949, Magma offered its shareholders rights to subscribe for 204,000 shares of capital stock of the company at $16.75 per share at the rate of one share for each two shares held. During the month preceding the offer, Magma had sold on the New York Stock Exchange at prices ranging between $19.75 and $16.50. The purpose of the offer was to raise another $3 million to finance continued underground exploratory development of the San Manuel mine, as a preliminary to a mine production project.

The issue was underwritten by Newmont and by Lazard Frères & Co., which had become interested in the project. As it turned out, Newmont and Lazard had to split equally the purchase of 98,917 Magma shares that remained unsubscribed. This, together with the purchase of the 30,400 shares offered to it as a Magma shareholder, brought Newmont's interest in Magma up to 22.9 per cent from 14.9 per cent.

Writing the technical portions of the prospectus for this issue was the first really responsible job Plato Malozemoff was given at Newmont since his employment in 1945. His work had, of course, to be cleared by the Securities and Exchange Commission, and here Malozemoff ran into an old friend of the Premium Price Plan days, Ben Adelstein, head of the technical mining section of S.E.C. Adelstein knew mining, having once worked with Anaconda, and despite his friendship with Malozemoff, or perhaps because of it, he seemed particularly strict and demanding in the process of clearing the Magma prospectus.

With the proceeds of the sale of new stock, Magma began sinking two shafts, one a permanent, the other an exploratory shaft, to a depth of about 2,000 feet, from which mine development work would be done and samples taken of the sulfide orebody for metallurgical test work. Meanwhile, efforts to obtain privately the main financing for the project were meeting stone walls everywhere, and the tension between McNab and Searls was increasing. Had McNab been willing to drop back to 10,000 tons per day, Newmont would have aided in the financing of the project. But McNab was obdurate, and it was 30,000 tons or nothing. Unless Searls and Newmont approved the project, it would always be nothing, as far as the private bankers were concerned.

How long this stalemate might have lasted cannot be guessed, but the Defense Production Act of 1950, an outgrowth of the Korean War, offered a way out. In October, 1950, James Boyd, director of the U.S. Bureau of Mines, called McNab and asked what needed to be done to bring San Manuel into production. Boyd thought help might be obtained under the act's provisions, and he asked McNab to come to Washington to talk things over.

It was the first of a great many trips to Washington "to talk things over." For the next two years, Magma's people were in and out of Washington constantly, going patiently and wearily from office to office and disappointment to disappointment. Yet, in the end, McNab and Magma won out by sheer refusal to give up.

On October 26, 1950, however, this was all hidden in the future as, following preliminary meetings which Roy Bonebrake also attended, McNab, Wes Goss, and E. T. Cummins, Magma's Washington representative, sat down with Jim Boyd to discuss San Manuel. Boyd said that the project did, indeed, come within the provisions of the Defense Production Act and that San Manuel should be brought into production as soon as possible. Boyd's direct, positive manner greatly encouraged McNab, and his dream of San Manuel began to firm up once more.

As Boyd outlined the procedure, it seemed fairly simple. The bureau approved of the idea, but bureau engineers would have to check Magma's cost estimates. Then Boyd and the General Services Administration (G.S.A.) would work out a copper procurement contract with Magma's representatives. Armed with this contract, Magma could then obtain private financing.

It was not all that simple.

In November, 1950, McNab sent to Boyd, as administrator of the Defense Minerals Administration (D.M.A.) a proposal regarding San Manuel. During November and December, bureau engineers checked the figures. During January and February, 1951, the terms of a procurement contract for San Manuel's copper were worked out in a long series of meetings between Magma, the D.M.A., and the G.S.A. On March 3, General Harrison of the Defense Production Administration (D.P.A.) told Jim Boyd that there was no problem and the certification would be issued shortly.

At that point, a freeze seemed to set in. Man after man in the D.P.A. or elsewhere got interested and wanted to review the contract, after which he would raise questions. Days and weeks went by, but every so often Magma would be assured that action would be taken very soon.

Then, in mid-April, D.P.A. decided to develop a "copper program" and naturally could not act on San Manuel until the program was detailed. More delays, and more assurances of prompt action. Finally, the program was ready, but then D.M.A. had no funds for it. The Congress had to act, which would take time.

Meanwhile, Boyd tried to work out an arrangement for a D.P.A. certificate to be issued "subject to availability of funds." On May 8, 1951, D.P.A. did issue such a certificate, and Roy Bonebrake and Cummins rushed it over to G.S.A. They needn't have hurried. G.S.A. didn't like the wording. D.P.A. should have said "would" instead of "intended."

Three days later, D.P.A. changed "intended" to "would," but added some other clauses that required more discussion to clarify. In the meantime, Congress had passed the legislation making additional borrowing authority available to the agencies to back up additional procurement contracts.

At long last, on May 15, 1951, Roy Bonebrake and a G.S.A. man worked out the last details of a copper procurement contract essentially the same as the one drawn up with D.M.A. and approved by D.P.A. Assured of the approval of Jess Larson, head of G.S.A., Roy signed it for Magma and went happily off to New York thinking he had a deal. He was mistaken.

Next day, Cummins went around to G.S.A. to pick up Magma's copy of the contract, signed by Jess Larson. To his dismay, he was told that Larson hadn't signed it and wasn't about to sign it. Nor could Cummins find out for days why Larson wouldn't sign or what Magma should do next. It was the run-around, all over again.

Finally, on May 23, Elliott, the G.S.A. general counsel, who had only just stepped into the matter, told Bonebrake and Cummins what G.S.A. wanted. Essentially, it was: to limit the project cost (for copper pricing purposes) to $70 million; to put a ceiling of $0.11 a pound on allowable production costs; to limit the minimum copper contract price to $0.21 a pound (the D.M.A. contract called for actual cost plus $0.07); to force Magma to sell a portion of its copper to G.S.A. at the minimum price even if the market price were higher; and a flat limitation of 420,000 tons of copper to be covered by the contract over a seven-year period.

Inasmuch as the project would certainly cost well over $90 million, and because the other terms were too severe, Magma had to reject the proposal. The rest of the summer and early fall went by in a tedious round of negotiations that did result in mid-November in a contract proposal that Magma thought it

LEFT: Blasting at O'okiep is done from long holes ring-drilled by machines like this one in Nababeep West mine. Driller Attie van den Heever supervises. RIGHT: O'okiep drives raises with this raise borer for increased efficiency and safety. Mine superintendent of Koperberg mine, Allan Dickson, checks operation.

In Tsumeb lead smelter (from left), James Ratledge, general manager; M. D. Banghart, retired managing director; and Dave Pearce, managing director; listen as Peter Philip, assistant general manager, makes a point.

Continual search for more occupies (from left), O'okiep's chief exploration geologist, Johan Marais; chief mine geologist, Dick van Zyl; and chief geologist, Andries Lombaard; shown here examining drill core at Koperberg Central.

O'okiep's copper smelter started with a nominal capacity of 14,000 tons of blister copper annually but with only minor enlargement now has a capacity of 50,000 tons per year. A second converter, shown here, was the only addition.

Magma in 1910 was based on this unimpressive layout, all that was left of the once-rich Silver Queen mine near Superior, Arizona, which W. B. Thompson bought for $130,000 cash.

Magma in 1972 embraces the San Manuel, Arizona, townsite, concentrator, smelter, and new electrolytic refinery and rod casting plant in foreground. Mine is several miles distant to right. *Ray Manley photo*

Magma's electrolytic copper refinery at San Manuel is newest and most modern in industry. Capacity is 200,000 tons of refined copper annually, much of which goes to consumers in Southwest and West Coast.

Continuous-cast copper rod is produced by this machine of Southwire design at San Manuel's refinery. Capacity is about 100,000 tons of rod annually in coils weighing up to 16,000 pounds each. *Ray Manley photos*

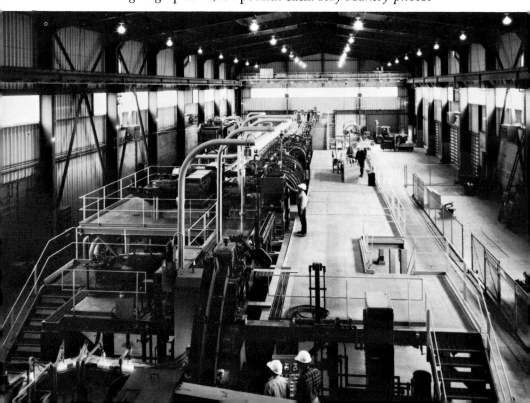

could live with. This was based on a mine output of 30,000 tons of ore a day, copper production of 70,000 tons per year, and a minimum price of $0.235 a pound, with no fixed set-aside at the minimum. Magma was to use this contract to obtain private financing.

It took only a few days of inquiry to confirm what McNab already feared. The banks and insurance companies had more loan requests than they had funds. Stuart Silloway told McNab that Mutual might grant a small loan if the contract were changed somewhat, but to get as much as $90 million he would have to include some thirty to forty participants in the loan, which would involve S.E.C. approval and would drag out the proceedings for months. Someone in Washington, perhaps Jim Boyd, advised McNab to go to the R.F.C.

McNab decided to do it, but first he had to surrender the G.S.A. agreement, the only tangible result of a year of negotiation in Washington. Back to G.S.A. it went, however, and McNab went off to the R.F.C. to tell his story all over again. This time, Magma insisted on a copper procurement contract to be negotiated with the Defense Materials Procurement Agency (D.M.P.A.), a relative newcomer, in order to back up repayment of the loan. This time, so much information was already on file that it was not too difficult to agree on a contract.

However, the R.F.C., enthusiastic at first, began to develop misgivings. Why wouldn't Newmont help finance San Manuel? The R.F.C. required an answer, so on May 2, 1952, Fred Searls wrote a letter to Roy Bonebrake, to be made a part of the record of the R.F.C. application, in which he told why Newmont was unwilling to finance the San Manuel project on the large scale desired by McNab.

In this letter, Searls said: "Newmont is willing to help Magma finance any plan of development of San Manuel that can be agreed upon between the two corporations. While uninformed people have assumed, and still state, that Magma is controlled by Newmont, we both know that such is not the case; and that Newmont, although owner of 22 per cent of Magma shares, has actually had no voice in Magma's plans and deliberations since 1948. Newmont has not been consulted, or even fully advised, as to the plans for financing and development of San Manuel, it being generally understood that a long standing difference of opinion has existed as to the size and details of the projected development.

"This letter is to reiterate the offer made orally at least three times over the past few years, to make available to Magma the time and services of myself and Mr. Malozemoff, together with such other technical assistance as may be required, to discuss and determine if in fact it is to the best interest of Magma shareholders to forthwith equip San Manuel on a basis of production of 10 million tons of ore per year."

The R.F.C.'s caution arose from the fact that it could not approve the San Manuel loan until it was satisfied that no private source, including Newmont, was willing to aid in the financing. However, on July 10, 1952, the R.F.C. did grant a loan to Magma of $94 million, the largest such loan it had yet made, for mine development and for construction of a mill and a smelter. The first disbursement of funds, $1,840,373.53, was made on February 21, 1953. The loan agreement included certain important conditions, such as requiring another stock offering, further injection of equity, construction of a townsite, and others. On August 29, 1952, Magma signed a procurement contract with D.M.P.A., and San Manuel was on the road at last.

The R.F.C. loan agreement required Magma to put in $6 million of its own money to be raised by sale of new common stock. In November, 1952, therefore, Magma offered one new share for each 2.375 shares of common already held at a price of $24.50 per share. A total of 266,227 shares was offered, and the issue was underwritten by Lazard Frères & Co. Fred Searls flatly refused to have Newmont underwrite this issue.

Newmont did purchase the 59,224 shares offered to it as a shareholder, maintaining its ownership of Magma unchanged at about 22 per cent. Lazard Frères, as underwriter, had to pick up 6,988 shares of unsubscribed stock, bringing its beneficial ownership to a little over 3 per cent. Not long previously, Lazard had owned about 16 per cent of Magma, but had sold most of its shares to various clients.

Credit should be given here to George Murnane, a partner in Lazard, who was of significant help in the entire San Manuel financing effort and who was a Magma director until his retirement in mid-1968. He died eight months later in February, 1969. He was a most articulate, astute, and persuasive financier, as well as a fine gentleman, a quality that marks so many of the men who appear in this history.

With the loan, the $6 million from the new stock issue, and the D.M.P.A. contract, San Manuel was in the clear—except for building the town of San Manuel, about a $12 million job. On November 15, 1952, after negotiation with several companies, Magma signed a contract with Del E. Webb Construction Co., to build 1,000 houses at San Manuel. Webb prepared a master plan for the town and detailed plans for the houses, all of which were examined and approved by Magma.

Webb then found that he could not obtain financing for the job except under Title IX of the National Housing Act, which required that the area be designated a critical defense area and that the Housing and Home Finance Agency (H.H.F.A.) set up the housing program. The nightmare began all over again.

At first, there seemed to be no problem. Then in March, 1953, the H.H.F.A. administrator said he thought the San Manuel townsite job must be opened for public bidding. Early in April the Federal Housing Administration (F.H.A.) office in Phoenix, on behalf of H.H.F.A., asked for bids on the San Manuel town job. Normally, the bids would have been opened and analyzed by the F.H.A. office in Phoenix, but late in April, the H.H.F.A. ordered the F.H.A. (Phoenix) to send all bids, sealed, to Washington.

On May 13, Messrs. Goss and Twitty, a Magma attorney, went to see Albert Cole, who had become H.H.F.A. administrator on March 10. Cole said that careful analysis by means of a special grid system had disclosed a tie between two bidders. Cole asked if Goss would split the job 50–50, but Goss quite properly declined. Cole then said that one of the two bidders in the tie was Del Webb and the other was MOW Homes, Inc. The latter company, which had no previous contact with Magma or San Manuel at any time, turned out to be a "paper company" with no tangible assets.

There followed a most amazing display of backing and filling by H.H.F.A., of argument by MOW, and attempts by Magma to follow some logical procedure. This went on all through May in almost daily meetings, but agreement was not reached until late at night on Friday, May 29, when the MOW backers agreed to allow Webb to be the prime contractor if MOW were provided a substantial interest in the San Manuel townsite job. It seemed the better part of valor to agree, and the parties spent all day Saturday working out the details. In the end it was done, and a creditable town was built at San Manuel. Magma eventually bought out both Webb and MOW and now operates the townsite itself.

In December, 1953, McNab was elected chairman of Magma and Wesley Goss was elected president. An unhappy event of that year was the death of Henry E. Dodge, who had been treasurer and a director of Magma since 1913. A year later, Walter P. Schmid was elected treasurer of Magma.

Walter Schmid has been associated with Newmont throughout its entire life. A native New Yorker, he began work with Newmont as an office boy in 1921 at the age of fourteen. An ambitious young man, he studied accounting at night school, eventually earning a college degree and his certificate as a Certified Public Accountant. As his abilities increased, so did his responsibilities in both Newmont and Magma, and Schmid eventually became controller of Newmont in 1956, and then treasurer in 1967. Schmid also became a director of Newmont in 1967 and has retained that post following his retirement in 1969.

A cheerful friendliness seems to be a Newmont characteristic, and in no one is it marked more strongly than in Walter Schmid. Yet, along with the ready smile, go a serious mind and a quiet determination to achieve a goal he believes to be worthwhile. Walter's analytical ability in finance and his persistence were both evident in the restructuring of Atlantic Cement, a Newmont affiliate, and Newmont Oil Company, a wholly owned subsidiary. These recent events were both of considerable significance to the parent company.

Engineering and construction for the San Manuel plant was contracted to a joint venture of Stearns-Roger Company of Denver and the Utah Construction Company of Salt Lake City. This work proceeded swiftly and by September, 1955, the 30,000-ton concentrator of San Manuel began operating on ore from development work. The smelter entered production in January, 1956. Also, thanks to careful planning by the Magma mining staff, the first ore block in the San Manuel mine began producing in the same month of 1956. Using a modification of the block caving method first worked out by Ohio Copper Company and Inspiration prior to 1910, San Manuel was up to production at a rate of 587,000 tons of ore per month within nine months after undercutting the first block.

The company's pleasure in this achievement was dimmed, however, on July 4, 1956, when A. J. McNab died suddenly in a hospital near his home in Pine Plains, New York. An officer and director since 1924, McNab had come to represent a quality of indomitability and perseverance that characterized the entire Magma operation, both at Superior and at San Manuel. The personification of integrity and truth, McNab never went back on a promise or forgot a friend. His secretary at that time, Miss Ruth Vanderpoel, who is now Plato Malozemoff's secretary, remembers McNab as "the strongest, the most modest, and kindest man I ever knew." A truly remarkable man, he was the leader of a truly remarkable enterprise.

In laying out the mine plan for the San Manuel orebody, the Magma engineering staff had the assistance of Arno Winther, a consulting engineer at the time for Miami Copper Company. Winther had been mine superintendent for various companies in the United States and in Rhodesia. By 1936 he had returned to the United States as manager of the Miami Copper Company's mine at Miami. Poor health required his temporary retirement in 1944, but upon recovering, he returned to work as a consultant, in which capacity he was engaged by McNab. Winther's specialty was block caving, a modification of which had been developed by Miami Copper in 1926 under the administration of F. W. Maclennan, who achieved considerable industry prominence for his development of this low-cost large-scale mining method. Unfortunately, Winther did not live to see San Manuel enter production, for he died of a heart attack in August, 1949.

The San Manuel staff had planned to support the various openings under each caving block of ore by means of heavy timber or steel sets. This type of support had previously proven sufficient in other mines in the industry using block caving. At San Manuel, however, owing to the great depth at which mining had to begin (1,400 feet), extremely high pressures developed rapidly in the first block, and the result was a great deal of expense, time, and manpower devoted to maintenance, particularly in the grizzly drifts and the haulage drifts under the caving block. So severe was this problem that it threatened the whole enterprise. After a long period of study and experiment, the difficulty was eliminated by substituting unreinforced concrete support for the timber. The result was a substantial reduction in maintenance costs and an increase in the efficient use of manpower. Although the problem was met and solved quite early in San Manuel's life, no mining problem since has presented quite so severe a challenge to the ingenuity of the Magma staff.

This difficulty in the caving operation, plus a few mechanical problems, held San Manuel's 1956 output down to a little over 78 million pounds of copper or 55.7 per cent of the planned capacity of 140 million pounds annually. However, Magma Copper Company's sales jumped from $18.8 million in 1955 to $43.8 million in 1956, due to a rising copper price and to the addition of San Manuel's production to that of the old Magma mine.

The rosy picture thus presented rapidly turned gray, however, when the price of copper began sliding in late 1956 and ended in 1957 at $0.245 per pound. Magma did sell some copper to the government at that price under the D.M.P.A. contract, but Magma's profit of $7,475,000 in 1956 melted to a loss of $2,340,000 for 1957. As if this were not enough, the company, which had barely struggled back into the black in 1958, ran into a strike in 1959 that shut down San Manuel for four months and Superior for over five months. The result was a loss for 1959 of $2,506,000.

There was nothing for it but to try to defer principal and interest payments due the Government under the R.F.C. loan. Roy Bonebrake, Wes Goss, and George Murnane went to Washington, explained the situation, and did obtain a reduction in principal payments in 1958 and a deferment of both principal and interest payments in 1959. By 1961, however, Magma was back on schedule, making payments even before they came due, and the average trend of Magma's earnings has been steadily upward ever since.

This is by no means the end of the Magma story. Actually, the fruition of Magma's growth potential developed in the late 1960's in a series of technologic and financial moves that greatly strengthened both Magma and Newmont. The purpose here has been only to describe the origins of Magma and particularly of San Manuel as a necessary preliminary to understanding one of the major forces that later made possible Newmont's phenomenal growth in the decade of the sixties. However, in the 1950's, faced with a multitude of fascinating opportunities, Newmont found itself in position to select the best of them and to develop each one to the fullest. The first such opportunity in the 1950's was the Sherritt Gordon venture, which must now be described because it marks such a significant point in the Newmont history and an even more critical event in the career of Plato Malozemoff, now Newmont's president and chairman.

Scraper hoist at Lynn Lake mine

10 Newmont Enters a New Era

Even knowledgeable people have been heard to say that in the timing and in the fact of acquiring a share in the O'okiep, the Tsumeb, the San Manuel, and the Sherritt Gordon enterprises, Newmont Mining Corporation "certainly was lucky." This statement could not be more untrue. "Lucky" mining companies usually wind up out of luck and out of business.

Involved in almost all successful mining operations have been one or two men who carefully weighed the odds, who overcame disagreement or outright opposition, and who found ways of coping with severe technological and financial problems. If they were also lucky, they had earned it, after all.

For Empire Star, the man was Fred Searls. For Tsumeb, it was Henry Smith and M. D. Banghart. For San Manuel, it was A. J. McNab and Wes Goss. For Sherritt Gordon, it was Eldon Brown and Malozemoff. The launching of each of these four enterprises involved considerable controversy within the company, and yet each of them contributed significantly, in its own way, to Newmont's progress.

Newmont entered the postwar world in the strongest position in its history up to that time. In 1945, Newmont Mining Corporation consisted of a group of extraordinarily talented individuals, whose abilities fortunately were complementary. The result of their activities, as we have seen, was a series of exploration, production, and financial enterprises, almost all of which were quite successful.

In the next few years, however, certain changes took place in the Newmont management structure that greatly influenced the company's future. In January, 1947, Judge Ayer retired as president and became chairman of the board. Fred Searls was elected president and Franz Schneider became executive vice president. This ended the existence of the triumvirate of Messrs. Searls, Schneider, and Ayer, which had managed the affairs of the company since 1931. After 1947 and until 1954, Fred Searls acted as chief executive officer of Newmont.

Also significant was that beginning at about this time, the Newmont board of directors acquired the services of a number of men, such as Justice Byrnes and General Lucius Clay, who had become quite friendly with Fred Searls during his work in Washington during the war. Byrnes, Clay, Lewis Douglas, and others who came after them, brought to the board's deliberations a useful insight into the changes taking place in the political and economic climate in which Newmont had thereafter to operate.

Still another, and even more significant, event of the time was the employment by Newmont in late 1945 of Plato Malozemoff, who in later years was responsible for some of the most profound and effective changes in Newmont's methods of management. To understand these changes, it is necessary to understand something of Malozemoff himself, of the atmosphere in which he grew up, and of his early experiences with the mining industry and with Newmont.

Born in Saint Petersburg, Russia, in 1909, Plato Malozemoff was the son of Alexander and Elizabeth Malozemoff, who were both members of the educated class, relatively small in Russia at the time. Education gave them something that was to them both an asset and a liability, namely a background that provided them the means of achievement well beyond the average level in Russia, but which also could easily bring them to the unwelcome attention of the Czar's secret police. Because of his activities in attempting to correct the many injustices that he saw around him, Plato's father was not only exiled to Siberia, he was imprisoned for a period on a charge of revolutionary activity.

As a result, the Malozemoff family settled in Siberia, several hundred miles northeast of Lake Baikal. Because of his engineering training, Alexander Malozemoff went to work for a gold mining company. Achieving rapid promotion, he was invited to become superintendent of a gold placer mine owned by the largest gold mining company in Russia, the Lena Goldfields Limited, of which he eventually became managing director.

Largely owned by British capital, Lena Goldfields was confiscated by the Soviets in 1920 following the revolution, and the Malozemoff family emigrated to the United States. Before long, however, Lenin realized that he would have to have outside help to rehabilitate the Russian economy, and he announced the New Economic Policy under which he invited foreign capital to help in Russian reconstruction. The senior Malozemoff returned to Russia in 1924 and negotiated a concession for Lena Goldfields on the confiscated properties in Siberia and also several others in the Ural and the Altai mountains.

Malozemoff also raised about $30 million from sources in America, England, and Germany and went to work to re-establish the mining enterprise. By 1929, Lena Goldfields was once more a going concern and a highly profitable one. At this point, Stalin reversed Lenin's policy, and once more Lena Goldfields was confiscated, this time permanently.

When the Soviets first seized the gold mines in 1920, Alexander Malozemoff was in New York, and Plato's mother, Elizabeth, received word from her husband to leave Siberia and join him. Travel to the west was blocked by sporadic fighting of the waning civil war, and the only alternative was to go east in the hope of rejoining Alexander at the Mongolian frontier or in Peking. Departing hurriedly, Elizabeth, Plato, and his brother, Andrew, set out on a journey that would have defeated anyone less resourceful and courageous than Elizabeth. Traveling by boat, in carriages, in wagons, by any means at all, after many months the family reached Verkhneudinsk (now Ulan-Ude), a town near the border of Outer Mongolia where they were detained under house arrest by the Cheka, the forerunner of the OGPU.

Unknown to Plato's mother, Alexander had already been at Maimachinsk (now Altanbulag) over in Mongolia, and after waiting for several weeks without news of his family, had given up and gone back to Peking.

Meanwhile, after a couple of months of worry and attempts to escape, Mrs. Malozemoff thought up a ruse that worked, and she and the two boys got safely across the border to Maimachinsk. Not finding her husband there, nor any news of him, Mrs. Malozemoff went on to Urga (now Ulan Bator, capital of Mongolia).

In Urga, the Malozemoffs ran into the so-called Chinese-Mongolian War, not a very big war, but big enough to block all means of further travel. This time it was strenuous effort, mostly argument, that won for Mrs. Malozemoff both permission to leave Urga and the means to do so. Here a real stroke of luck was involved, for the family learned later that on the very next day after they had left, the town of Urga was invaded by Baron von Sternberg, a deranged former officer of the Imperial Russian Army who had turned bandit. Picturing himself as a savior of the yellow race, von Sternberg slaughtered all the Russians in Urga, and proclaimed himself the Emperor of Mongolia. Several months later, von Sternberg set out to conquer Siberia as well, and his empire and his life ended shortly thereafter before a Red Army firing squad.

In Peking, the Malozemoff family was reunited and continued its flight to San Francisco, where the family settled. There Plato pursued his interest in the violin, inspired by his father who was a talented violinst. At one time, Plato considered making a career as an orchestral musician, particularly during the Depression when no engineering jobs were to be had.

In due course Plato entered the University of California at Berkeley. Because of his familiarity with his father's profession, Plato studied mining engineering, graduating in 1931 with a B.S. *magna cum laude*. He then spent a year at the Montana School of Mines in Butte, Montana, receiving a master's degree in metallurgical engineering, also *cum laude*. Next came two additional years in Butte where he worked in the laboratory of the State Department of Mines, carried on studies in mineral dressing under Professor A. M. Gaudin, and in addition, worked nights for the W.P.A. in Butte.

Those Depression years have a special significance for Plato Malozemoff in that from about 1930 on, doing part-time work, he was the sole support of his family. The senior Malozemoff had put practically all of his earnings from Lena Goldfields into U.S. and British stocks. Between the stock market crash of 1929 and the Soviet confiscation of Lena Goldfields that year, he lost everything. Therefore, while his father was re-establishing himself, Plato's income was all that the family had. With no one to turn to and nowhere to go for help, life became a struggle for existence. As it did on everyone who lived through it, the Depression left its mark on Malozemoff in the form of an increased ambition, a never-say-die determination, a new self-reliance, and a keen cost-consciousness. These qualities strongly mark Newmont today, just as they characterize its president.

After leaving Butte in 1934, Plato spent a summer in France, then returned to California, joining the Pan American Engineering Company in Berkeley, which was a partnership doing laboratory work in the field of mineral dressing and also manufacturing flotation machines, jigs, and other mill equipment. Malozemoff's starting salary at Pan American was the princely (for 1934) sum of $150 per month.

After some four years with Pan American, Malozemoff joined his father in managing a small gold-copper mine in Argentina. Technically, the mine and mill were well established, but the ore reserve turned out to be insufficient to support the operation. From Argentina, Plato and his father turned to another gold mine in Costa Rica, and although the property appeared to have interesting prospects, it was doomed by the wartime difficulties imposed on gold mines generally.

When the Costa Rican venture passed into the hands of a Canadian mining group, Plato and his father returned to the United States in 1943, where Plato found employment in the Minerals Branch of the Office of Price Administration.

The Minerals Branch of O.P.A. and the War Production Board (W.P.B.) jointly administered the Premium Price Plan, and the Metals Reserve Company financed it. The intent of the Premium Price Plan was to obtain an increase in the production of copper, lead, and zinc beyond those amounts that might have been expected or rendered possible under the market prices of the three metals, which, of course, were fixed under the O.P.A. The Premium Price Plan assigned price payments, in addition to the controlled metal prices, for specified tonnages of metals to be produced by certain mines. Mining companies considering themselves eligible would apply to the Premium Price Plan for this special help, and based upon their analysis of the facts surrounding the mining company's operations, the engineers on the staff of the Premium Price Plan would then grant or deny special quotas and prices to each company that applied. Malozemoff's work on this staff was outstanding.

Although in his work Malozemoff analyzed probably hundreds of such applications, he was given the chief responsibility for all of Newmont's properties. Among these was Resurrection Mining Company in Leadville, Colorado, where he made a thorough analysis in a remarkably short time, ending with a recommendation for assistance. Malozemoff's report on Resurrection greatly impressed Phil Kraft, who concluded that Malozemoff was just the sort of younger man that Newmont should bring into the management. In addition, A. J. McNab and Henry Smith were both familiar with Malozemoff's work in the Premium Price Plan and were favorably impressed by it.

When Kraft and McNab approached Fred Searls with the idea, however, they found him unenthusiastic. Although Phil and others arranged meetings between Searls and Malozemoff, Fred remained unimpressed for about a year, during which time Malozemoff received other offers, which he was reluctant to accept. In the end, Searls agreed to give the young man a try, and Malozemoff joined Newmont on October 1, 1945.

Coincidence or not, this date marks the start of a period of tremendous growth in Newmont, both in terms of assets and of earnings. For example, in 1946, Newmont reported the fair market value of its assets as $69.2 million. In 1971, it was over $900 million. In 1946, Newmont's earnings were $2,467,269. In 1971, the company earned $54,520,000. Obviously, Malozemoff is only one of many men responsible for such progress, but it is a fact nevertheless that Newmont's greatest growth has taken place under his leadership.

When Malozemoff began work with Newmont in New York in 1945, he found that his new associates had a fairly clear idea as to what the future of the company should be. For Newmont at least, gold mining had lost the attractiveness it had possessed in the early 1930's. Inexorably rising costs were squeezing profits against gold's ceiling price, which made any future profit clearly dependent on government edict rather than the free play of supply and demand. Therefore precious metals were sidetracked, and base metal exploration and development had again taken the lead.

Listed on the payroll as staff mining engineer, Malozemoff soon discovered that he was in fact all the engineering staff there was. Franz Schneider delighted in introducing Plato as "Newmont Mining Corporation's staff." Additionally, Phil Kraft introduced Plato to the Canadian exploration activities that he was directing. As time went on, Malozemoff was called upon by Henry Smith, Franz Schneider, Henry Dodge, A. J. McNab, in fact everyone in the office, for assistance on various exploration, metallurgical, or financial projects as the older members developed confidence in his ability.

Finally, Fred Searls, who had been rather distant at first, became aware that the mining company analyses prepared by Malozemoff were both thorough and accurate, and Fred began turning over to Malozemoff a few prospects that had been submitted to Newmont for possible investment. The result was that before long, Searls channeled all such submissions to Malozemoff. For years thereafter, Searls would only proceed further with a proposition if Malozemoff had been over it first and recommended it. Searls was very much his own man still, however, and did not always follow Malozemoff's recommendation.

As Searls became more and more impressed with Malozemoff's ability, he devised another kind of trial to see what the new staff member could do. This took the form of turning over to Malozemoff a portfolio of stocks that Searls, himself, had been handling for some time as a part of his activity for helping to finance exploration by way of capital gains from what he called the "Wall Street Stope." Never one for half measures, Searls simply turned over to Malozemoff about $360,000 in common stocks with the injunction, "See what you can do with it." From 1947 to 1958, Malozemoff worked the "Wall Street Stope" and on an average investment of $303,000 earned capital gains of $298,000, plus $184,000 in dividends for the eleven-year period. Since 1958, this sideline has in turn been a testing ground for still younger men who came into Newmont after Malozemoff.

For several years, Plato Malozemoff occupied himself diligently as "the Newmont staff." Except for assisting on projects such as Tsumeb and on the S.E.C. registration statements used for the initial financing of San Manuel, no really large project came Malozemoff's way until the Sherritt Gordon enterprise was brought to Newmont's attention in 1950. Because the Sherritt Gordon development was the first in a series of new enterprises much larger in scope than Newmont had yet attempted, it seems appropriate to tell the story here in some detail.

It must be said at the outset, however, that although the
Sherritt development was remarkably successful technologically,
it was somewhat disappointing financially. Only in recent years
has Sherritt begun to fulfill the high hopes originally held for its
financial success.

Sherritt Gordon Mines Limited owes its existence to the
discovery of a copper-zinc deposit in Manitoba, made shortly after
the First World War by a Cree Indian prospector, Phillip Sherlett. Carlton W. Sherritt, a twenty-three-year-old trapper working in the same area, staked some claims adjoining Sherlett's.
Unfortunately for Sherlett, his agent failed to record the assessment work at the recording office in The Pas. When the claims
went open, Sherritt and his partner restaked Sherlett's claims.

Sherritt then optioned the property to J. P. Gordon, who
was scouting in the area on behalf of his two brothers, who were
lumbermen in Ontario. Not long after, Sherritt died in an air-
plane crash, and J. P. Gordon and his brothers optioned the entire property to three different companies in succession, but unsuccessfully. Finally Bob Jowsey, a prominent Canadian mining
man, got them together with Thayer Lindsley, and Sherritt Gordon Mines Limited was formed in 1927 as a result. Phillip Sherlett
did not lose out entirely, however, for Eldon L. Brown, who became manager and eventually president of Sherritt Gordon, suggested an annual payment to Sherlett and his wife, which the
company made as long as the two lived.

Sherritt Gordon Mines Limited had been in production for
only a year or so when the copper market hit bottom in 1932.
The mine closed down and the town of Sherridon, named for
the two founders, became deserted. A rising copper market later
in the 1930's brought about the reopening of the mine and the
resurrection of the town. The first superintendent and manager
of the Sherridon mine, Eldon L. Brown, returned to manage the
property, and before ore reserves ran out in 1951, the mine had
yielded about $59 million worth of copper and other metals.

Brown had not waited idly for Sherridon's ore reserves to
disappear. He had sent several prospectors into the field, and

one of them, a man named Austin McVeigh, found what he was looking for. After five years of roaming through northern Manitoba, McVeigh came on an outcrop that looked interesting enough to sample. As such things sometimes happen, he found this outcrop more or less by accident, and from it there developed the Lynn Lake mine. The very outcrop still exists in the center of a complex of headframe and other buildings at Lynn Lake. The samples McVeigh took from his outcrop assayed 1.5 per cent nickel, and 1 per cent copper, sufficiently attractive to warrant further study.

McVeigh made his discovery in the summer of 1941, but owing to shortages of men and supplies, nothing could be done about it until after the war. Having kept the discovery a close secret, Eldon Brown sent McVeigh back to Lynn Lake in 1945 to stake out claims. This time, the discovery could not be kept secret and a prospecting rush took place, but McVeigh had found the best locations.

Drilling the Lynn Lake orebody and others discovered later nearby aggregated more than 14 million tons of ore averaging about 1.2 per cent nickel and 0.62 per cent copper. Metallurgical test work, including the operation of a 25-ton pilot flotation plant at Lynn Lake, indicated that concentration was no great problem. However, a roadblock developed when it came to considering disposal of the nickel and copper concentrates. Shipment to the nearest nickel smelter in eastern Canada would eat up part of the probable profits and the smelting costs more than took care of the rest. In other words, Lynn Lake was a fine orebody and yielded ore that could be concentrated easily, but without a cheap metallurgical procedure to produce the actual metals, the operation could go nowhere.

At this point, there began an exciting story of the combined efforts of several metallurgical and financial talents. No one of these talents would have been sufficient to solve the problem alone, but working together, they turned Sherritt Gordon from a rather dubious prospect into a viable mining and metallurgical operation. The first step was taken by Eldon Brown in

1947, when he engaged Professor Frank A. Forward of the University of British Columbia to test on the Sherritt Gordon ore a technique Forward had developed for recovering metal by hydrometallurgy instead of by smelting. In about a year of work, Professor Forward succeeded in developing a process for using an ammonia leach on the sulfide concentrates from Lynn Lake and to recover from the leach solution the ore's nickel, copper, and cobalt.

At about the same time, the Chemical Construction Company in New York had developed a hydrogen precipitation technique that fitted in remarkably well with Forward's discovery. Brown got the two of them together, and plans were laid for a pilot plant to test the combined procedures. Charles Hames was at that time manager of Sherritt's Metallurgical Research Division. Vladimir Mackiw, a competent European chemist, who had emigrated to Canada after the war, directed the bench-scale development of the process. Preliminary test work indicated the possibility of achieving a remarkably successful and low-cost operation, but outside financing was clearly needed.

After several unsuccessful attempts to finance the venture through Canadian mining companies, Eldon Brown mentioned this problem to John Drybrough, his brother-in-law. In turn, Drybrough went to Newmont, with which he had a close relationship developed in several Canadian gold mining ventures. Drybrough's first contact at Newmont was with Malozemoff, who listened with interest until Drybrough mentioned that the Sherritt Gordon process depended on leaching with ammonia. Because of the rather limited past usefulness of ammonia in copper leaching, Malozemoff's first reaction was negative. Drybrough, however, urged him not to sell Eldon Brown short, because Drybrough had found Brown to be a man of great judgment and ability. Somewhat reluctantly, Malozemoff agreed to talk to Brown, whereupon Brown came down to New York and spent an entire day with Malozemoff describing the Sherritt Gordon process and what the company planned to do with it.

Little by little, Malozemoff's interest grew. For one thing,

Sherritt Gordon was not gold mining. For another, as Brown described it, the process involved a new and quite interesting metallurgical technique. Finally, it was the first opportunity Malozemoff had had to study on his own a really promising and large-scale proposition.

The result was that Malozemoff began a thorough investigation of Sherritt Gordon from the mine right up through the products of the proposed refinery. Before long, Malozemoff had become convinced that the Lynn Lake prospect of Sherritt Gordon had not only substantial ore reserves, but a process that might be made to turn out nickel, copper, and cobalt economically and profitably at prices then in effect.

Malozemoff also learned that Eldon Brown, before he came to Newmont, had been told gratuitously by a well-known engineering consultant that no one had yet developed a successful hydrometallurgical process for handling nickel and copper sulfides, and Sherritt wasn't likely to do so, either. Also, two Canadian mining companies had been told of the project and had declined to participate. With a Searls-like independence of thought, Malozemoff dismissed these opinions as having been based on inadequate analyses and continued his study.

Drybrough remembers being in Fred Searls's office one day and watching Malozemoff walk up and down in front of Searls's desk, keeping up a steady stream of argument, which was so convincing that Searls finally relented and promised to have Newmont back the project.

For several years thereafter, Malozemoff spent at least 90 per cent of his time on the Sherritt Gordon development. During all this period, he had the realization that for him, at least, the Sherritt Gordon venture was most critical. It was apparent to him, although not stated, that the outcome would have a great effect on his position with Newmont one way or the other.

With a pilot plant in operation at Ottawa to test the new process, Malozemoff turned to the formidable problem of financing the venture. The total cost of equipping the mine, the concentrator, and the metallurgical plant would come to around

$47 million. Brown calculated that about $11 million could come out of Sherritt Gordon's earnings. Another $4 million could be obtained by increasing Sherritt Gordon's capitalization through offering new shares at the rate of $2.00 a share. Another $5 million could be borrowed. This left about $27 million to be raised by outside financing.

In this effort, Malozemoff was fortunate to have the assistance of Franz Schneider, who had been so effective for Newmont during the difficult years of the Depression. It was Schneider's opinion that among Newmont's friends in the financial community, it should be possible to develop sufficient funds to take care of the entire venture. Foremost among these friends, was J. P. Morgan & Co., Inc.

With Schneider backing him up, Malozemoff was able to convince J. P. Morgan & Co. that the Sherritt Gordon venture had real merit. The Morgan Bank then went to work and put together a group of banks and insurance companies, led by Metropolitan Life Insurance Company, that was willing to supply the funds, provided Malozemoff could satisfy Metropolitan that the new process had a better-than-even chance of success.

Walter Page of Morgan led these efforts, later joined by Elmore Patterson. Also of assistance at this time was Stuart Silloway of The Mutual Insurance Company of New York, who brought his company into the group. Silloway, as has been seen, also considered helping finance the San Manuel development.

All in all, about $19 million in loans was arranged in this manner, and in addition Newmont itself eventually bought a total of $8 million worth of Sherritt Gordon convertible debentures. In addition, in 1951, Newmont bought 1,122,196 shares of previously unissued Sherritt common at $2.00 (Canadian) per share. To anticipate somewhat, it might be mentioned that by May, 1968, all Sherritt's obligations in this financing had been paid off as they came due, and Newmont converted its debentures into common stock of Sherritt Gordon. Newmont now owns 39.4 per cent of Sherritt Gordon Mines Limited.

Malozemoff's attention now turned to the Sherritt Gordon

pilot plant, which had completed its first major run in 1950. Not satisfied with the results of the first run, Malozemoff realized that the close scrutiny he wished to give the second major pilot plant test (officially designated as Run 23 and started in November, 1951) would require more time and attention than he had available. What Malozemoff needed was an independent engineer who could represent him during pilot plant tests. Accordingly, Malozemoff looked about for a chemical or metallurgical engineer possessing an analytical and logical mind, and preferably one who would be neither dismayed nor carried away by the new approaches of the Sherritt process. Such a man would have to check every possible source of error, and he would have to withhold his approval stubbornly until all such checks had been made to his satisfaction. Malozemoff found such a man in the person of Oscar F. Tangel, at that time assistant head of the Mineral Dressing Division of Battelle Memorial Institute in Columbus, Ohio.

This was not Malozemoff's first contact with Tangel. Having graduated from Lafayette College with a B.S. in mining engineering, Tangel began in September, 1933, a year of study on a fellowship at the Montana School of Mines in Butte, under Professor A. M. Gaudin. This was Malozemoff's last year in Butte, and he and Tangel became well acquainted through working in the same laboratory during the year. After receiving his master's degree in metallurgical engineering in June, 1934, Tangel spent the next school year at the Freiberg Bergakademie, Saxony, Germany, as a Fellow of the Institute of International Education. After his return to the United States in the fall of 1935, he worked for a number of rather weird and ill-fated small gold mining enterprises in California and Nevada, and finally hit a more satisfactory connection with the Fresnillo Mining Company in Mexico, where he became assistant mill superintendent. During the war, Tangel served as a captain in the Office of Chief of Ordnance, United States Army, and after the war he returned to Battelle where he had spent three months before Pearl Harbor. In 1966, Tangel left Battelle and joined Newmont in

New York where he is now vice president of research and development.

Tangel remembers the Sherritt Gordon job as one of his toughest assignments while at Battelle, in that it required continual analysis and evaluation of work involving technical problems that seemed insurmountable, and complicated chemical reactions that were far from being well understood. Tangel was fortunate in being able to have the assistance of various specialists drawn from the Battelle staff, especially James F. Shea, research engineer.

Tangel believes that a good part of the success of the overall Sherritt process must be credited to Eldon Brown, who created and maintained a determined team spirit among the rather heterogeneous metallurgical staff, made up as it was of Professor Forward, representatives of Chemical Construction Company, and of Sherritt Gordon itself.

For example, Tangel remembers that after his first meeting with the Sherritt staff, Eldon Brown turned to him and said, "Look here, Mr. Tangel, if you fellows turn in a report saying that the process won't work, I will just not believe it."

"That's all right, Mr. Brown," replied Tangel. "We'll just report the facts and give you our opinions separately." Later, Tangel came to realize that Brown had not been expressing any distrust of Battelle. On the other hand, he had simply stated his complete and utter conviction that the Sherritt Gordon staff would make the process work. Oscar saw that Brown had developed an unshakable faith in his staff that left no room for even the slightest doubt and, in fact, provided for no option other than complete success. Of course the staff knew that Brown believed in them, and Tangel thinks they worked much harder than they might have if they had had to fight management's doubts along with the technological problems.

Briefly, the original Sherritt Gordon process consists of several successive steps as follows: The sulfide concentrate from Lynn Lake, originally carrying about 12 per cent nickel, 1.5 per cent copper, and 0.7 per cent cobalt, but now somewhat lower in grade, is leached with ammonia at a temperature of 175 degrees Fahrenheit and a pressure of 105 psig in autoclaves. After thickening and filtration, the filtrate is boiled to precipitate most of the copper, and the remainder of the copper is stripped out by autoclaving with hydrogen sulfide at 250 degrees Fahrenheit and 15 psig.

The next step, which is the production of pure nickel powder, is the one that caused most of the trouble in developing the process. Although nickel powder can be produced from the pregnant, copper-free leach solution, traces of unsaturated sulfur compounds in the solution would lead to contamination of the nickel powder and the ammonium sulfate fertilizer.

After extensive research, during which the Sherritt Gordon staff had to develop a completely new body of knowledge, a processing step was developed which Sherritt Gordon calls "oxydrolysis." In the first stage of oxydrolysis, essentially complete oxidation of thiosulfate and polythionate is achieved by autoclaving at 350 degrees Fahrenheit and 700 psig. The temperature is then raised above 450 degrees Fahrenheit and the autoclaving is continued until all sulfamates are hydrolized to sulfates. Subsequent operating experience led to a reduction in the number of autoclaves required by adjusting conditions so that a significant amount of the oxidation and hydrolysis takes place in the heat exchangers employed in this circuit.

After oxydrolysis, nickel is reduced from solution by hydrogen at 350 degrees Fahrenheit and 500 psig total pressure in an autoclave. The reduced mother liquor, containing about equal amounts of nickel and cobalt, is treated with hydrogen sulfide at 150 degrees Fahrenheit and a pressure of 5 psig. The solution contains ammonium sulfate, which is crystallized and sold as a fertilizer. The nickel-cobalt sulfide precipitate is leached and treated to recover the contained metals.

Almost as difficult as the chemical problems indicated in the foregoing summary were the mechanical problems. Means had to be developed for transferring sometimes highly abrasive pulps to and from autoclaves that were operating under pressures as high as 700 psig. This meant development of seals around shafts and valves associated with the autoclaves. However, during the second pilot plant run, such mechanical problems were handled on a miniature scale or in ways that could not have been scaled up for commercial operation. By the end of this second pilot plant run in November, 1952, the staff had a much better idea of the chemistry of the process and felt quite confident of eventual success.

However, long before the end of the run, doubts had been raised because the concentrate fed to the pilot plant had been oxidized by long storage, so much so that about 10 per cent of the nickel content was water-soluble. In addition, ammonia losses were unexplainably high and metallurgical balances were not entirely satisfactory.

All this had greatly concerned Malozemoff, and his doubts grew as the run progressed. He called a meeting of all interested parties in late May, 1952, in Judge Ayer's office in New York to discuss still another pilot plant test. Eldon Brown and Vladimir Mackiw of Sherritt Gordon, Ed Roberts and Sid Nashner of Chemico were all for proceeding without a third pilot plant test on the ground that what bugs remained in the process could be worked out in a commercial plant. Malozemoff insisted that a third run on fresh concentrate would be required. In a long and sometimes heated discussion of the advisability of further work, all the elements of an additional pilot plant run were thoroughly thrashed out.

In the end, it came down to Malozemoff's saying flatly that he simply could not agree that enough work had been done, and until he, himself, was satisfied he would be unable to satisfy the lenders. Until the lenders were satisfied, there would be no money forthcoming for commercial plant construction.

Accordingly, a third major pilot plant run (Run 24) was agreed upon in June, 1952, well before the completion of the second test. In this final run, some of the chemical fundamentals of the process were to be studied further, and details of operating techniques were to be developed. Thus at a cost of some six to eight months, and about $1 million, the additional pilot plant run was completed to the satisfaction of all concerned. Sherritt Gordon received the funds and speeded up construction of the new plant and refinery at Fort Saskatchewan, which had actually been started in June, 1952.

In addition, a new town had to be built at Lynn Lake along with a concentrator. It occurred to Eldon Brown that he had a ready-made town and concentrator at Sherridon, where the mine was now worked out. In a development that has no parallel anywhere in the mining industry, Sherritt Gordon literally packed up 208 buildings, including a bank, a post office, a school, and two churches, as well as 150 houses, and moved them along with the concentrator 200 miles up to Lynn Lake. Because the intervening muskeg was impassable in the summertime, all moving had to be done in the dead of winter, with tractors hauling the buildings on sledges.

When financing was completed, the Sherritt Gordon board was reconstituted with the addition of Messrs. Malozemoff, Drybrough, Bonebrake, and Schneider to the membership. Any disagreement over the pilot plant was quickly forgotten as the successful test proceeded, and from that time on, Newmont and Sherritt Gordon have worked most cooperatively.

However, about a year and a half after construction started, it developed that about $8 million more in financing was going to be required than had been estimated. With the completion of construction still about eight months away, money was running short. A re-estimation of construction costs indicated that there would be a 40 to 50 per cent overrun.

Malozemoff checked the figures, took a deep breath, and told the people at the Morgan Bank what was going to happen. After the initial shock wore off, the bank's reply was that Newmont would simply have to put up whatever additional money was needed.

Henry Alexander, chairman of the Morgan Bank, called a meeting to discuss the situation. Malozemoff presented the facts of the Sherritt Gordon venture at that meeting, and also presented an analysis indicating that despite the overrun, the venture was still a good bet. In this opinion, he was backed up by Fred Searls, who by now had become completely convinced that the whole enterprise had real merit. The problem was eventually solved by obtaining an advance of $5 million from the General Services Administration for nickel to be produced from Sherritt's stockpile, and by the agreement of Newmont to buy an additional $1 million of subordinated debentures and $205,000 in bonds, and of the Morgan Bank to buy another $2,795,000 in Sherritt bonds.

On the road once more, construction proceeded rapidly, and the first nickel powder was produced on July 21, 1954. In the meantime, the Canadian government had agreed that the Canadian National Railway would build a line into Lynn Lake at a cost of about $15 million. On November 9, 1953, Donald Gordon, president of Canadian National, drove a nickel spike at Lynn Lake to mark the completion of the rail line. The first car of concentrate started that day on its way to the refinery at Fort Saskatchewan.

For the first few years after the Lynn Lake mine and the refinery entered production, the high grade of the "EL" orebody, (about 2.5 per cent nickel), which was mined first, yielded profits of around $6 million to $9 million annually. Thereafter, Sherritt entered a long period of declining earnings and did not pay a dividend until 1960. In recent years, however, operations were improved by the start of a new high-grade copper-zinc mine, the Fox mine, and, in addition, higher metal prices greatly enhanced Sherritt's performance. For example, in 1970 the company's net income was $15,843,000 (Canadian funds), but lower metal prices in 1971 cut the net to $7,497,000.

Largely owing to the thoroughness of the pilot plant work, the complicated Sherritt Gordon process experienced a minimum of start-up difficulties. As the result of a fairly steady expansion program, the refinery in 1971 attained a production of 33,111,000 pounds of nickel metal, 3,618,000 pounds of refined copper sulphides, and 561,000 pounds of cobalt, a good deal of it from metal scrap as well as from Lynn Lake concentrates.

In addition, the refinery is the center of a chemical complex which, through purchase of phosphate rock from Florida and by manufacture of sulfuric acid, is able to offer large tonnage quantities of a wide range of fertilizers in addition to the by-product ammonium sulfate produced in the refinery. Sherritt's laboratory has also expanded to include pilot plant testing of nickel ores from all over the world, including, especially, the low-grade lateritic nickel ores from the South Pacific. Most recently, the Fort Saskatchewan pilot plant has tested samples from the lateritic ores of P. T. Pacific Nikkel in Indonesia (in which Newmont and Sherritt have a 15 per cent and a 10 per cent interest, respectively) and those of Société le Nickel in New Caledonia. Following a similar test program, licenses for the Sherritt Gordon process were issued to Marinduque Mining and Industrial Corporation in the Philippines for a plant on Nonoc Island that will have a capacity of 75 million pounds of nickel annually. Another licensee is Western Mining Corporation Limited which has built a nickel refinery in Western Australia. Union Corporation of South Australia also has in operation a small nickel refinery licensed by Sherritt Gordon.

For the future, Sherritt is looking to two new copper mines in Manitoba that were discovered since 1968. The Fox Mine, worked by underground methods, started production in 1970 and had, at the end of 1971, reserves of about 14.5 million tons assaying 2.35 per cent zinc and 1.99 per cent copper. Another orebody, at Ruttan Lake, has been partially drilled to reveal about 51 million tons of ore averaging 1.61 per cent zinc and 1.47 per cent copper. This orebody lies close to the surface and can be exploited initially by open-pit methods. In 1971, the property was being equipped for production by mid-1973 at a rate of 10,000 tons of ore per day.

The Sherritt Gordon venture has special significance in the Newmont history for several reasons. Some of them are obvious, such as the profitable outcome of the venture against formidable technological obstacles, and the good relationship established with a growing Canadian enterprise. Less obvious, but of far greater significance, was the fact that success with Sherritt Gordon initiated a change in the Newmont thought processes.

Prior to the Sherritt venture, in which Newmont invested about $13 million, Newmont's management shuddered at the thought of risking more than a million or two million dollars on a new property. True, Magma did successfully launch San Manuel as a $94 million project in 1951, but Newmont put little money into it.

After Sherritt Gordon, Newmont began to think in larger terms. The company's executives and its board, gingerly at first, but eventually quite calmly, became able to consider propositions calling for raising sums they once would have been unwilling even to mention.

For example, Newmont's investments of $13 million in Southern Peru, $11.9 million in Palabora, $33 million in Atlantic Cement, $67.5 million in Granduc, and, as yet, a not fully determined but substantial investment in Magma, all these, and more, lay in Newmont's future. Sherritt really opened the company's eyes to its own great potential.

With the story of Sherritt, therefore, the Newmont history enters one of its most exciting periods, the crowded, furiously busy, often argumentative, sometimes disappointing, frequently rewarding decade that began about 1954, the year when Plato Malozemoff became president of Newmont.

Granduc's mill at Tide Lake, B. C.

11 New Philosophy in Action

It was as if a set of springs had been compressed behind New-mont Mining Corporation during the 1940's, and success with Sherritt Gordon had released them. In earnings, assets, accomplishments, the company shot ahead in the 1950's. For one thing, the Sherritt Gordon experience had developed Malozemoff's confidence in himself and in the technical and financial power of Newmont. For another, Newmont now had real confidence in Malozemoff. Following Sherritt Gordon, the Newmont enterprises increased in number, they increased greatly in size and in financing required, and, most important, increased sharply in asset value to the parent company. Working together in earnest, the Newmont management in the three years following the third and successful pilot plant run at Sherritt Gordon in 1952, helped organize, finance, and launch seven major mining projects and two major Newmont investments. The largest in tonnage was Southern Peru. The most interesting in its variety of minerals was Palabora. The knottiest problem was presented by Granduc.

Of these nine projects, five, Palabora, Dawn (uranium), Cassiar Asbestos, Newmont Overseas Petroleum Company, and Southern Peru Copper Company, were outstandingly successful, although adverse political developments have hampered the last two projects. One project, Atlas Consolidated Mining and Development Company in the Philippines, was for a few years quite profitable in net income and capital gains to Newmont. Another venture, Western Nickel, about broke even. One project, Granduc, has yet to prove itself. Only one project, the smallest, which was an attempt to smelt nickel ore in Baja California, Mexico, by a new process, was a flat failure.

Not a bad record for only three years of activity, particularly when one remembers that, simultaneously, adequate attention had to be given to Tsumeb, O'okiep, the North African lead-zinc properties, and the domestic operations of Magma, Idarado, Resurrection, and Empire Star.

However, the rapid multiplication of Newmont's interests, spreading as they did over a good part of the free world, generated certain accounting and reporting problems. Some of these difficulties stemmed from the fact that in June, 1942, Newmont Mining Corporation was required by the Securities and Exchange Commission to register as an investment company under the terms of the Investment Company Act of 1940. That Newmont deserved this designation at that time emanated from the activities of Colonel Thompson, who, prior to the establishment of Newmont, had devoted himself to promoting various mining companies, bringing them to a point of profitability, and then selling them, in whole or in part, to someone else.

As we have seen, however, Thompson obviously did not intend Newmont to continue in this fashion and certainly since the Depression, Newmont depended less and less on its investment trading activities and more and more on its income from its managed mining properties. Following Thompson, Judge Ayer continued the practice of maintaining a portfolio of stocks, many of them for investment purposes, but he greatly reduced the number of such holdings from the multiplicity of Thompson's portfolio. After Ayer, Fred Searls went still further in the direction of reducing the number of investments and of regarding them more as liquid sources of funds for use in developing and equipping properties that his exploration activities had turned up.

Nevertheless, Newmont still, in 1940, had not escaped identification as an investment company, albeit widely recognized as a remarkably successful one. After studying the Investment Company Act of 1940, Judge Ayer, Fred Searls, Fred's brother Carroll, and Roy Bonebrake drew up a tentative application for exemption from registering as an investment company under the act. This application was presented informally to the S.E.C., and quite as informally, it was rejected. Under the Investment Company Act, if a company's primary activity was in the business of investing, reinvesting, owning, holding, or trading in securities, and if in addition, the company had over 40 per cent of its total assets in the form of securities other than government securities, or those of majority-owned subsidiaries of the company, it was considered by the S.E.C. as an investment company under the act.

Although in the Newmont charter it is clearly stated that the company is to be actively engaged in the business of exploring for, acquiring, financing, and operating mining properties, the critical point was that in 1942, of Newmont's assets in security holdings of about $38 million, only 22 per cent was in the securities of controlled companies, and 78 per cent consisted of miscellaneous investments in companies not controlled by Newmont, which was far in excess of the S.E.C.'s 40 per cent limit. For the purposes of the Investment Company Act, a company was considered to be presumptively controlled if the parent company owned over 25 per cent of its securities.

Therefore, in 1942 the S.E.C. classified Newmont as a "closed end non-diversified investment company of the management type." This was not particularly burdensome to the com-

pany during the 1940's, but with the upsurge of activity in the
early 1950's, the investment company status began to be galling.
As an investment company, Newmont had to file detailed annual and quarterly statements on a special form that was much
more complicated than the Form 10-K that other mining companies filed with the S.E.C. In addition, there were a number of
special restrictions that covered such things as sale or purchase
of the company's own stock, or that required S.E.C. approval of
many types of deals, which took so long that the deal might be
lost before approval was granted. Also, an investment company
cannot file a consolidated financial statement except in consolidation with other investment companies, although Newmont
had no interest at that time in filing a consolidated statement.
Finally, it was by no means absolutely clear that, as an investment company, Newmont might not have to obtain stockholder
approval every time it invested in a mining venture of a speculative nature. After all, what mining ventures are not speculative
to some extent?

Therefore, in early 1955, Newmont petitioned the S.E.C.
for exemption from the Investment Company Act pointing out
that as of the end of 1954 Newmont had 54 per cent of its assets
in companies owned or controlled, and only 46 per cent of its assets in companies not wholly owned or controlled, and all of
these were mining or petroleum or exploration companies.
In addition, Newmont pointed out that 90 per cent of its officers'
and staff members' time was devoted entirely to mining activities.
Furthermore, Newmont did not bear even a superficial resemblance to an investment company, inasmuch as it had no investment committee and it employed no security analyst. Whereas in
1935 Newmont held thirty-two different common stock issues
listed on various exchanges, by the end of 1954 this had been reduced to thirteen such holdings. In that period, Newmont's net
assets went from $45.8 million in 1935 to over $200 million at
the end of 1954. A final point was that in 1954, 8,000 employees
of companies controlled or managed by Newmont produced
41,000 tons of copper, 58,000 tons of lead, 39,000 tons of zinc,

P. Malozemoff joined Newmont as staff engineer in 1945, became president in 1954 and also chairman in 1966. Newmont's greatest growth has occurred during his stewardship. *Fabian Bachrach*

Roy C. Bonebrake came to Newmont in 1939 from Stanford Law School. Active in both Newmont and Magma, he became a Newmont vice president in 1954, and general counsel and a board member in 1959. In 1965 he was elected executive vice president of Newmont, retiring on December 31, 1971. *Guy Gillette photo*

LEFT: Philip Kraft (left) joined Newmont as a geologist in 1925, devoting himself to Canadian exploration and eventually to Newmont Oil Company, of which he was president until 1955, and chairman until 1968. Dr. Robert S. Moehlman (right), oil geologist, became president of Newmont Oil in 1962. RIGHT: Marcus D. Banghart started at Newmont's Island Mountain mine in British Columbia in 1934. He served thereafter at Berens River and O'okiep, becoming Newmont's vice president, operations, in 1954. A board member since 1955, he retired in May, 1971. *Guy Gillette photos*

Hal Norman (left) has served Newmont all over the world as exploration geologist since 1950. Robert J. Searls (right), Fred Searls's son, is a Newmont vice president and managing director of Newmont Proprietary Limited.

LEFT: Frank W. McQuiston, Jr., was involved in nearly all of Newmont's metallurgical projects since 1934. He retired as vice president in 1969, is now a consultant.   RIGHT: Fred Scheck, with Newmont since 1931, has handled accounting and project evaluation problems. He retired as secretary of O'okiep and Tsumeb in 1970. *Guy Gillette photos*

Signing the Magma-R.F.C. loan agreement for San Manuel in 1953 were: (l. to r., seated) Hector C. Haight, R.F.C.; Wesley P. Goss, Magma; A. J. McNab, Magma; (standing) Howard Twitty, Magma attorney; Roy C. Bonebrake, Magma; John Ingram, R.F.C.; A. E. DeCelles, R.F.C.

LEFT: Jacques L. Leroy (left), Newmont's international legal counsel, joined the company in 1955. Jack E. Thompson (right), executive vice president, started with Newmont in 1960 in new project development.
RIGHT: Eugene H. Tucker, chief engineer and a forty-one-year veteran with Newmont, has worked at, or visited, about every Newmont location around the world. *Guy Gillette photos*

LEFT: William T. Smith (left) and Walter P. Schmid, two generations of treasurers, were Newmont employees from 1927 and 1921, respectively. Smith retired in 1967, Schmid, in 1969.   RIGHT: David O. Pearce, Newmont's vice president, operations, and a Newmont director, joined O'okiep as mine superintendent in 1955, became general manager in 1959, and is now managing director O'okiep and Tsumeb.

Guy Gillette photo

57,000 ounces of gold, and 1.3 million ounces of silver. None of this could have been done by a true investment company.

Following a hearing in Washington in October, 1955, the S.E.C. issued an order specifically exempting Newmont from the classification of an investment company under the Investment Company Act of 1940. The S.E.C. order clearly stated that Newmont was a company primarily engaged in the business of exploring for, acquiring, and financing mining and petroleum properties in keeping with the company's charter.

The reason for explaining this development in some detail is the hope that it may serve to clear up a misconception that still exists among some members of the financial community. One of the more widely read financial advisory services still recently referred to Newmont Mining Corporation as "essentially a closed-end investment trust specializing in the shares of nonferrous metals and oil producers." Newmont is nothing of the sort, as this chronicle should have made clear by this time.

With its position as a mining company officially established, although not as yet apparently widely recognized, Newmont proceeded rapidly with the development of the many new enterprises previously mentioned. For Malozemoff himself, promotion within the company followed rapidly on the heels of his successful performance with the Sherritt Gordon project. He became a Newmont vice president in December, 1952, a director of Newmont in May, 1953, and in January, 1954, he was elected president and chief executive officer. Fred Searls became chairman of the board at that time. Henry DeWitt Smith and Franz Schneider, who had been towers of strength for the company in their respective fields, retired as officers at the end of 1954, but remained as directors. Remaining or becoming officers in 1954 were M. D. Banghart in charge of operations, Roy C. Bonebrake and Carroll Searls handling legal matters, and William T. Smith and Walter P. Schmid handling the financial affairs of the company. Phil Kraft was also a vice president at that time, devoting practically all his time and attention to the affairs of Newmont Oil Company.

Thus, Searls and Malozemoff had the continued assistance of three of the stalwarts of the previous twenty years, plus the talents of a new and younger group of officers. In addition, Newmont was no longer dependent on a one-man staff as it was when Malozemoff joined the company. In the mid-fifties, there were, in New York, Bob Fulton in exploration, Frank McQuiston in metallurgy (just back from four years with the Atomic Energy Commission), Gene Tucker as chief engineer, Jack Mann on special projects mostly related to California, and Henry Volkman brought from Resurrection to assist in purchasing. At the newly constructed laboratory in Danbury, Connecticut, there were L. J. Bechaud, Jr., in charge of metallurgical research, and Dr. Arthur A. Brant, in charge of geophysical research and development. Obviously, Newmont was growing away from the independent pillar type of structure that had characterized it during the thirties and forties, into a more closely knit organization where a number of individual talents could be brought to bear on any particular problem. Opportunities for testing this newly developed organization multiplied rapidly in the early 1950's.

One of the most interesting such projects because of its ramifications was Western Nickel. Located in British Columbia near the town of Hope, Western Nickel drew Fred Searls's attention shortly after Newmont became involved with Sherritt Gordon. This copper-nickel orebody had been explored to some extent by a small company called Pacific Nickel Mines Limited. In 1952, the Defense Minerals Exploration Agency (D.M.E.A.) was offering special inducements for discovery and development of nickel orebodies, owing to a shortage of the metal. Searls and Pacific Nickel organized Western Nickel Limited, in which ownership was shared 51 per cent Newmont and 49 per cent Pacific Nickel. Under Searls's direction, the company set up an exploration program in 1952.

To explore Western Nickel's pipelike sulfide orebody at a deeper level, Newmont drove a mile-long tunnel and did a good deal of underground development work in addition to about 10,000 feet of diamond drilling. This work cost about

$700,000 but developed only about 170,000 tons of additional ore for a total reserve of about 1,162,000 tons assaying 1.36 per cent nickel and 0.56 per cent copper. This did not seem sufficiently attractive to justify putting the mine into production, and the project was therefore closed down in October, 1954.

About two years later, nickel prices in Europe began increasing substantially, and with each jump in the nickel price, Western Nickel looked more and more attractive. Also by 1956, Sherritt Gordon had demonstrated the value of its new refining technique, and it appeared that nickel concentrate from Western Nickel could be refined by Sherritt Gordon at a good profit based on prices prevailing in European markets and in merchant markets in America. At the very least, by reactivating Western Nickel, Newmont saw an opportunity to recover the earlier exploration expense.

To be sure of a market, Newmont entered into negotiation in Europe with potential customers located by Fred Jeffrey, of International Minerals and Metals, agents for Newmont. The product was to be refined nickel briquettes produced by Sherritt Gordon. Three European customers expressed willingness to sign contracts for the nickel briquettes at specific tonnages and set prices of around $1.60 a pound. These were: Klöckner & Company in Germany; Schoeller-Blechmann in Austria, and Fiat in Italy.

With these contracts in hand, Malozemoff, now Newmont's president, set about financing the project. To begin with, the shareholders of Western Nickel, which in 1956 included Newmont (46.1%), Sherritt Gordon (4.9%), and Pacific Nickel (49.0%), advanced a part of the funds required for construction. Newmont's share of this was $195,650. In addition, Western Nickel obtained, through Malozemoff's assistance, a loan of $2,825,000 from New York banks. This loan was guaranteed by the shareholders in proportion to the amount of their ownership. Newmont's share of this guarantee was 46.1 per cent. This was the first time since the Depression that Newmont had borrowed money or guaranteed a loan by a subsidiary. The burden of the

$3 million debt to the Morgan Bank in the early 1930's had been so distasteful to Newmont that the company shunned further borrowing until this time.

Granby Consolidated Mining, Smelting & Power Company, Limited, a British Columbia copper producer, contracted to manage the project and, to the great delight of Fred Searls, who doted on such economies, Granby aided in mill construction by contributing at a low appraisal value a great many items of used mill equipment, such as ball mills, pumps, classifiers, and flotation machines. Construction began in July, 1957, and was finished at the relatively low cost of $880,000 for a capacity of 1,000 tons of ore per day. The mill began operating in January, 1958.

Although quite well conceived, the Western Nickel venture was unfortunately timed in that the nickel market began to weaken almost simultaneously with the beginning of nickel shipments to Europe. Early in 1958, nickel prices became so low that the three European customers declined to accept shipments under their fixed-price contracts. Following lengthy negotiation conducted by John Drybrough and Jacques Leroy, the sales contracts were partially cancelled, and each of the three customers made payments to Western Nickel in lieu of accepting further shipments at the contract prices. The original contracts of May 9, 1957, were cancelled in August, 1958, by means of these cash settlements.

Apparently, the United States Department of State and, particularly, the U.S. Embassy in Austria were of considerable assistance in reaching this settlement. The influence of the Embassy in Austria was of particular importance because the Austrian government was a major partner in the firm of Schoeller-Blechmann.

Before operations at Western Nickel came to an end, the company did repay all of its obligations, and a small profit was left for the partners, although one of them commented at the time that "nobody made any money out of this except Sherritt Gordon." Sherritt Gordon, of course, did the refining.

Western Nickel was dissolved in July, 1960, and some of the equipment was sold. The property and the mill were sold to Giant Nickel Limited, and operations were resumed in July, 1959. Until March, 1960, all production went to Sherritt Gordon, but thereafter concentrate was sent to Sumitomo in Japan under a contract that has been extended year by year. Mill capacity was increased to 1,400 tons per day in 1962. Nickel production has amounted to about 4 million pounds annually, together with 2 million pounds of copper. Giant Nickel is owned by Giant Mascot Mines Limited of Vancouver, British Columbia. This venture is of some importance in Newmont's history because of the sad experience with the fixed-price nickel purchase contracts, and because of the experience of guaranteeing a loan for a subsidiary, neither of which have been much to Newmont's liking.

The second, and vastly more significant, development that began in the early 1950's was that of Granduc Mines Limited. One of the most fascinating aspects of the history of mine development is the fact, demonstrated over and over again, that enormous mining enterprises can grow out of a single casual conversation. As we have seen, this was true of San Manuel, where the whole enterprise really began with a conversation overheard by John Gustafson. In the case of Granduc Mines Limited, this large copper project grew out of a conversation between John Drybrough and Don Cannon, a geologist Drybrough had employed in 1952. When Cannon got back from his first field trip, Drybrough asked him what he had seen that looked interesting. Cannon described the Granduc prospect in northern British Columbia, which was being investigated by Granby Consolidated, Newmont's associate in Western Nickel.

The name, Granduc, was made up by combining Granby, the developers of the property, and LeDuc, from the name of the river and the glacier located nearby. From the very beginning of the project, the natural obstacles, even to exploration, have been unusually formidable even in an industry that has had to tackle almost every imaginable difficulty of nature.

For many years, the only really convenient way of reaching the Granduc property was by air. Nevertheless, a prospector named Einar Kvale hiked over the mountains to the Leduc area in 1948 and staked some claims. Three years later, he was back, this time working for Karl Springer's Helicopter Exploration Company Limited, a pioneer in the use of helicopters for mineral exploration. The party staked a good many claims that year and in 1952, all on behalf of Springer's company.

Granby heard of the staking and in 1952 sent in a crew to see what was going on. The reports so encouraged Granby that the company worked out a deal with Helicopter Exploration that led in March, 1953, to the formation of Granduc Mines, Limited, the present owner of the property.

Although relatively high in copper content, the Granduc orebody was so situated as to discourage any number of prospective investors that Granby attempted to interest in the property. Guarded by rugged, high mountains on the one hand, and the LeDuc Glacier on the other, there seemed no way into the property except by a lengthy tunnel, and no one wanted to undertake the enormous expense of driving such a tunnel.

Despite all this, Cannon was enthusiastic about Granduc, and catching some of this enthusiasm, Drybrough wrote to Larry Postle of Granby Consolidated in 1952 to say that if Granby needed additional funds to develop the Granduc prospect, Newmont might be interested in supplying them. A year later, after some of his other possibilities had washed out, Postle came to see Drybrough and asked if Newmont would like to buy half of Granby's position in Granduc, which Newmont, after some investigation, did in December, 1953.

An ambitious exploration program was then set up under the direction of Larry Postle, including sinking a shaft and development on a lower level. He continued in charge of the project until February, 1959. At that time, the obviously high cost of development and the rather small size of the orebodies disclosed by the drilling discouraged the partners, and the project was shut down, although not abandoned.

During the shutdown, Malozemoff and Drybrough asked Hal Norman, a Newmont geologist, to study the area in the hope of finding more ore. Norman's report on his studies was a classic of careful and beautifully reasoned geological deduction that explained many of Granduc's mysteries. Although Norman found no "outside" orebodies, he was able to demonstrate by a small amount of drilling, the existence on the other side of the glacier of a high-grade counterpart of the known orebody. Norman also postulated an extension of the known orebody under the glacier, and together, these two discoveries opened the possibility of a large enough addition to ore reserves to warrant the resumption of exploration at Granduc.

In 1962, the partners resumed the surface and underground drilling program, and it was completed by 1963. At this point, there had been indicated approximately 32.5 million tons of ore assaying 1.93 per cent copper. Although around $6 million had been spent on the project by this time, and a great deal of study had gone into solving the natural problems involved in mining the ore, no easy way of handling the job had been developed. It looked, in 1963, as if around $55 million in new capital would be required to get the property into production.

Although Fred Searls had never been keen on the Granduc project at all, Malozemoff's imagination had been fired by the challenge presented by Granduc. At this point, Hecla Mining Company, of Wallace, Idaho, approached Malozemoff and, after some discussion, expressed a willingness to purchase around $10 million worth of Granduc common and preferred shares. In 1964, Malozemoff then approached American Smelting and Refining Company and found Asarco willing to provide $10 million as advance payment on copper concentrate that would eventually be produced by Granduc and would be shipped to the Asarco smelter at Tacoma, Washington. A $30 million bank loan commitment was also secured, and Newmont was to provide the remainder of the $55 million estimated capital requirement.

With financing at least temporarily provided for, work

began at both ends of the 10.3 mile access tunnel to connect the mining operations with the concentrator, which was to be located down at Tide Lake, on the other side of a 7,000-foot mountain range at the foot of the Berendon glacier. From Tide Lake to the nearest port at Stewart is 31.5 miles by road, which required building 20 miles of new road alongside Salmon glacier and improving the existing road for the rest of the way. Added to this factor of isolation, forbidding climatic conditions, including an unusually heavy snowfall, make it extremely difficult to build a permanent camp in the open in the neighborhood of the mine proper.

Things have not been easy for Granduc, particularly its weather problems. The first winter was unusually severe. More than 90 feet of snow fell on the work camp at the mine. In February, 1965, an avalanche roared down on the work camp killing twenty-six men, injuring nineteen others, and wrecking a good part of the camp and the installations at the mine and tunnel entrance. Mervyn Upham, who was at that time a Newmont vice president in charge of Granduc, reached the mine immediately and was most effective in directing rescue operations.

Although a thorough government investigation cleared Granduc Mines Limited of any negligence or of any violation of safety precautions in establishing the camp where it did, everyone in the company was profoundly shocked by what had happened, almost to the point of giving up the project. To make sure there would be no repetition of this accident, Malozemoff secured a world-famous snow avalanche expert, Montgomery Atwater, to analyze the causes of the avalanche and to recommend measures for the avoidance of future avalanche damage. In addition to studying avalanche control, engineering contracting firms prepared new detailed estimates, brought up to date, of construction and equipment costs.

The result was the clear indication that owing to inflated wage scales and equipment and supply costs, the expenditure on Granduc would be much higher than had been anticipated when the original estimates were made. Furthermore, it proved to be considerably more expensive than expected to move men and materials from Stewart up to the two construction camps. In addition, there had been, apparently, a number of rather serious underestimates made because of the preliminary nature of the engineering work. Finally, the avoidance of future avalanche damage required expensive inspection procedures by experts, bombing snow slopes from a helicopter, and shooting down snow accumulation with 75 mm artillery pieces that were permanently emplaced at several stations along the road. In short, Malozemoff had to face an expenditure of at least $30 million higher than the original $55 million estimate.

Granduc Mines Limited now had a critical financial problem as well as the technical problems, which had been severe enough. Since June, 1964, Newmont and Hecla had been financing much of the work by advances to Granduc, which by the end of September, 1965, had aggregated $5.6 million in addition to the $10 million originally put in by Hecla for purchase of Granduc stock.

Newmont at this point had a 45.8 per cent interest in Granduc Mines Limited. With the nearly 55 per cent indicated overrun, meeting the increase by bank borrowings was doubtful. The many public shareholders could not be counted on for the additional money, and it appeared that the principal burden of providing additional funds would fall on Newmont, even though it owned less than 50 per cent of the company. Furthermore, the profit return for a company burdened with this large depreciable capital investment would not be taxable in Canada for many years, resulting in the dividends paid by Granduc Mines Limited to Newmont becoming fully taxable in the United States at the then 52 per cent rate under the laws of the United States without any benefit of foreign tax credit, depreciation, depletion, or tax-deductible interest.

Several unsuccessful attempts were made to interest different mining companies in Canada in becoming shareholders in the company (Canadian companies were approached because they would avoid the tax disadvantages of a United States company such as Newmont).

A solution was eventually worked out by Malozemoff and his legal associates, Roy Bonebrake and Frank Rinehart, aided by Walter Schmid, that was accepted by all parties. First, Newmont formed a wholly owned U.S. subsidiary called Granduc Operating Company. That company and American Smelting and Refining Company agreed to lease in equal shares the property of Granduc Mines Limited, to proceed with the development and equipping of the mine for production of approximately 7,000 tons of ore per day, and to provide all the necessary funds. At the same time, Newmont Mining Corporation surrendered the shares that it owned of Granduc Mines Limited to that company for cancellation. In return, Granduc Mines Limited as lessor was to receive a royalty equal to 22.5 per cent of the net profits (before depreciation and financial charges, but after current capital charges) from production until 32.5 million tons of the ore reserves were milled, and 25 per cent thereafter. Newmont's subsidiary, Granduc Operating Company, was to be the manager of the operation.

To shareholders of Granduc Mines Limited, the lease agreement offered a guaranteed entry into production and relieved them of the necessity to provide any additional funds, and through the royalties, they could hope for a sizable income. Cancellation of Newmont's shares decreased the capitalization of the Granduc company with a proportionate increase in the percentage ownership of the remaining shareholders. Thus Hecla, for example, had its ownership increased from 16.67 per cent to 34.5 per cent of the outstanding shares of Granduc Mines Limited.

For Newmont, through its wholly owned U.S. subsidiary, Granduc Operating Company, and for American Smelting and Refining, the lease agreement provided a way of entering production with certain important tax advantages. Such a leasing agreement permitted Newmont and Asarco each to write off its share of the more than $40 million development expense against their own profits from elsewhere and to deduct property taxes, depreciation, and depletion before imposition of U.S. Federal income taxes on profits.

In the meantime, work on the property had continued as fast as possible, although still harassed by natural and man-made problems. Ore reserves were recalculated using a lower grade and different cut-off points in the light of the expected ore extraction methods to be used. The 1966 figure, before waste dilution, was 43,343,000 tons assaying 1.73 per cent copper. Work in the tunnel, which was now being advanced from the Tide Lake side, was proceeding swiftly, occasionally obtaining a rate as high as 93 feet a day. Driving the tunnel was never a simple matter. A two-month delay after the tunnel was in about 5,000 feet was caused by a heavy flow of high-pressure water. Grouting ahead of the tunnel finally sealed off this flow. Still later, at between 13,000 and 17,500 feet, weak graphitic shale rock required extensive installation of steel arch support and concreting, which delayed progress for another three months. Despite all this, record-setting advances were achieved here and there, including at one point, a one-mile advance in only 73 working days. The tunnel was completed in December, 1968.

After a string of heartbreaking delays and disappointments, Granduc's mill began rolling on November 1, 1970, and the first copper concentrates went off to Japan in January, 1971. Ore production gradually increased in 1971, reaching a rate of between 5,000 and 6,000 tons per day in late 1971. However, great difficulties were experienced in starting production. Because of Granduc's remoteness and the unsettled labor conditions arising from the impending termination of Granduc's labor contract in May, 1971, the labor turnover was extremely high, on the order of 35 per cent a month. Under such conditions, training labor in the series of complex operations required at Granduc proved most difficult and results were disappointing. Ore dilution in mining by the sublevel caving method used was excessive, as had been experienced in the early stages by other mines using this method. Also, productivity was low and costs consequently were high. The Japanese smelters that bought the concentrates raised their charges and limited their purchases in order to cope with their own problems. All these factors, coupled with a drop in copper prices in 1971, resulted in a financial loss for Granduc Operating Company in 1971 of $4,621,000 including charges for depreciation, depletion, and other non-cash items amounting to $2,430,000.

Although the labor situation had improved greatly by the end of 1971, much remains to be done in increasing productivity and reducing both mining costs and dilution of ore in mining. Only in this way can Granduc be set on a profitable path, and only the future can tell to what extent the monumental effort to bring this enterprise to realization will have been worthwhile.

A healthy sort of restlessness is a Newmont characteristic, and seldom has this been more graphically demonstrated than in the mid-1950's. Along about the same time that Newmont was developing an interest in the Granduc project, an opportunity arose for the company to enter a quite different field, that of uranium production. This came about through Fred Searls's friendship with Spencer Hinsdale, the Portland banker who, with Oscar Johnson, had been instrumental in bringing Newmont into the Idarado project.

Along about 1949, two members of the Spokane Indian Tribe, twin brothers named Jim and John LeBrett, were out looking for tungsten in one of the mountainous areas of the Spokane Reservation, located in the state of Washington. Their black-light lamp did locate a fluorescent mineral, but it turned out to be autunite, a hydrous calcium-uranium phosphate. This identification was made by a friend, Robert Hundhausen, who had once been employed by the Bureau of Mines. He encouraged the LeBrett brothers to look further.

As more evidence of uranium mineralization developed, the LeBretts went for help to their uncle, Clare Wyncoop, who owned a ranch on the reservation and had considerable business experience. With his help the Indians secured a mining lease on their discovery, organized a company, Midnite Mines Incorporated, and began exploring the likeliest uranium showings. In this they were guided by Hundhausen who had a business associate named Spencer Hinsdale.

In time the work indicated the presence of a sizeable deposit of open pit uranium ore, and Midnite began to do some mining. The group actually shipped some ore to Vitro Chemical Company's plant in Salt Lake City, but it looked to Hinsdale and Hundhausen as if the Indians needed more help. Hinsdale, who knew Fred Searls through his interest in Idarado, telephoned him and asked if Newmont would be interested in assisting Midnite develop its property.

Searls's interest was aroused at once, and following an examination of the property, Searls organized an exploration and metallurgical campaign. Diamond drilling indicated substantial quantities of ore high enough in grade in uranium to respond readily to metallurgical treatment. Newmont's metallurgical staff developed a suitable method of concentration and Kilbourn Engineering Company of Canada estimated that the mine could be developed and a concentrator built for about $6 million.

It appeared that the company could obtain a purchase contract with the Atomic Energy Commission on the basis of which the original investment could be paid off by 1960, and a profitable operation indicated for at least five years thereafter. Therefore, in 1956, Newmont organized a new company, of which, Midnite Mines Incorporated owned 49 per cent and Newmont owned 51 per cent. The new company was called Dawn Mining Company, perhaps in poetic recognition of a more favorable outlook than that of "Midnite." Newmont also loaned Dawn $1,620,000 to get mill construction started. By August, 1957, under the direction of Bob Fulton, a Newmont geologist and engineer who was manager at Dawn from 1956 to 1958, the 400-ton-per-day mill had been completed at a cost less than the original estimate. By October, the mill was operating at full capacity.

Dawn began to repay a bank loan of $3 million in December, 1957, in fact the company had shown a profit for the period of August through December of that year.

In 1958, Dawn paid Newmont its first dividend of $51,000. Thereafter, Dawn averaged over $1 million in dividends to Newmont each year until the A.E.C. contract expired in 1965. Lacking further contracts, Dawn was closed down and the mill put on standby.

Dawn again began producing in January, 1968, when Bob Fulton negotiated an agreement to sell uranium oxide to the Jersey Central Power & Light Company and to the Metropolitan Edison Company, both owners of large atomic energy plants. The contracts called for a minimum of 2.5 million pounds of uranium oxide to be delivered over a period of three and a half years, beginning January 1, 1970.

In June and July of 1951, Newmont looked over several uranium properties in the Beaverlodge area of Saskatchewan, near where Eldorado Mining and Refining, Ltd., had discovered pitchblende and was planning to build a mill. Two of these were the Rix Athabaska and the Amax Athabaska, which were controlled by Joe Hirshhorn, who eventually made a fortune in Canadian uranium. The examination was also to include a neighboring claim called the Clix. Incidentally, this particular Amax had no relationship whatsoever to American Metal Climax, Inc.

Following a visit by Dick Murphy of Newmont in June, 1951, John Drybrough went up to have a look. On July 13, Drybrough sent Malozemoff a telegram that is a Newmont classic for brevity and point.

AMAX CLIX AND RIX ALL NIX. JOHN DRYBROUGH

On August 10, 1951, Malozemoff received a nine page, single-spaced letter from Dick Murphy on these and two other uranium prospects making the same point as the telegram, but in detailed geological terms. It must be mentioned, however, that Rix Athabaska was developed by others into a small, short-lived uranium mine that was in production from 1954 to March, 1960.

Newmont could hardly afford to devote as much time and effort to every enterprise in which it invested as it did to Dawn, Granduc, and Sherritt Gordon. At about the same time Newmont became involved with these companies, it also invested in other enterprises where much less management participation was required. A good example of the latter type was an opportunity that arose in 1953 when John Drybrough became interested in the operations of Cassiar Asbestos Company, which had just started production from a very high-grade mine in northern British Columbia.

Fred Connell, the head of Cassiar, was a friend of Drybrough's, and in conversation one day in 1952, Drybrough asked Connell if Cassiar might have any interest in selling some of its stock to Newmont. Connell replied that he might be interested, and a couple of days later, he called Drybrough to offer 150,000 shares of Cassiar to Newmont at $6.00 a share. Newmont accepted, and a year later, bought an additional 500,000 shares of Cassiar from another Canadian company, Con West, at $8.00 a share. Still later, Newmont purchased additional shares, and now owns about a 13 per cent interest in Cassiar Asbestos.

Purchase of the Cassiar shares was a kind of diversification for Newmont in that it did involve a degree of management participation, although not to the extent required by several other properties. Cassiar operates two asbestos mines, one just south of the Yukon border in British Columbia, and the other 40 miles west of Dawson, Yukon Territory. These two properties produce a high-quality, long-fiber asbestos. On a scale of 1 to 10 in decreasing quality, Cassiar's asbestos grades from No. 1 to No. 5. Most asbestos produced in Canada is graded Nos. 5, 6, or 7. One of Cassiar's mines has about thirty years of ore reserves, but the other has a much shorter life in prospect. At present, Drybrough and Malozemoff represent Newmont on the Cassiar board.

This brief description of a few of the projects in which Newmont became involved in the furiously active years of the 1950's can give only a partial idea of the electric, sometimes stormy, but never dull, atmosphere prevailing in the Newmont office in that decade. Literally hundreds of prospects were brought to the company, and all of them had to be examined and analyzed in greater or less detail.

Dozens and dozens of metallurgical tests had to be made on various ores, reports had to be written and carefully studied. Probably the most fascinating aspect of the work of a mining company such as Newmont is that each new venture presents a different set of challenges. The basic elements are the same in that they involve mining, treating, and selling mineral products, but there is infinite variation possible on this basic theme. For Sherritt Gordon, the challenge was metallurgical. For Granduc, for a time the greater challenge was financial, but is now a mining one. For Western Nickel, the principal problem was markets.

Each new project launched, each new challenge faced and overcome, simply increased the Newmont appetite for more of the same. Newmont was steadily growing in reputation, in technological skill, and, thanks largely to O'okiep and Tsumeb, in financial resources as well. The growing strength of Malozemoff's leadership set a bolder course for Newmont, leading into larger and more complex enterprises in the foreign field than Newmont had heretofore dared consider.

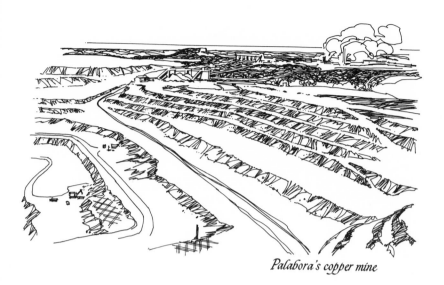

Palabora's copper mine

12 Launching Ventures Abroad

Excessive caution is not a characteristic that makes for success in the mining industry. The greater the prize, the greater the risks a mining company is willing to assume in order to grasp for it. Thus in the mid-1950's, even though the Newmont management was aware that political developments in various parts of the world were making foreign mining enterprises less and less attractive, certain opportunities developed in 1955 and 1956 that simply could not be turned down without displaying an extremely uncharacteristic and excessive caution.

Four such outstanding opportunities were those of Atlas Consolidated Mining and Development Company in the Philippines, Southern Peru Copper Corporation in Peru, Palabora Mining Company Limited in South Africa, and Newmont Overseas Petroleum Company in Algeria. All four presented interesting challenges, all four were outstandingly successful, but only Palabora has escaped political interference.

Because it was a petroleum venture, the rise and fall of Newmont Overseas Petroleum Company is described in Chapter 15. The other three projects appear in the following pages.

As happens so frequently, Newmont's interest in Atlas Consolidated came about through a casual conversation. Sometime in 1956 Plato Malozemoff had lunch with Clem Pollock, in charge of exploration for American Smelting and Refining Company. Just back from the Philippines, Pollock had been impressed by the possibilities for finding additional copper ore reserves at the property of Atlas Consolidated on the island of Cebu, although not sufficiently to cause him to recommend the property to his own company. After further investigation, Malozemoff had Newmont begin buying Atlas stock. At about the same time Colonel Andrés Soriano, the chief executive of Atlas, was in the United States looking for financial and technical help in developing the Atlas property.

Organized in 1953, Atlas mined its first copper ore in 1955. Ore from the company's Lutopan pit ran somewhat better than 1 per cent copper, and enabled the company to operate profitably through 1960, with a low of $510,000 in profits for 1958 and a high of $885,000 in 1959. Not satisfied with this rather meager initial showing, Colonel Soriano knew the company needed more financing and more technical help.

One of the men Colonel Soriano approached in New York was Philip D. Wilson of Lehman Brothers. Wilson was quite close to Newmont, having been manager at O'okiep during the exploration days in the late 1920's. Wilson suggested a conversation with Malozemoff, who by that time had built Newmont's position in Atlas to about 60,000 shares at an average cost of about $24 per share. Apparently the two men hit it off at once, and Colonel Soriano was particularly impressed by the fact that even without examining the property, Malozemoff had enough faith in it to invest rather heavily in it.

Colonel Soriano agreed to have a Newmont group examine the Atlas property. To save time and to achieve thoroughness, Newmont sent Fred Searls, Fred's son, Robert J. Searls, Richard

M. Belliveau, an open-pit mining consultant, and M. D. Banghart to do the job. The group spent the two weeks between December 5 and 19, 1956, at Atlas's Lutopan property. Based on
their studies, these men came back with the recommendation
that ore reserves be checked, additional reserves be sought, and
that open-pit operations be improved and expanded to increase
profitability.

Malozemoff and Soriano then proceeded to work out in
1957 a most unusual and imaginative series of agreements
whereby 242,000 block shares (10 Philippine shares equalled one
block share) were set aside for Newmont Mining Corporation to
be delivered at the rate of 1,512.5 shares for every 100,000 tons
of new ore added to the Atlas ore reserves as a result of a New-
mont-supervised geological study. In addition, Newmont would
receive a 10 per cent royalty interest on all ore production from
such new orebodies. However, Newmont could exchange this 10
per cent interest for 151,250 shares. Also, Newmont could acquire a 15 per cent royalty interest by finding 6 million tons,
which would yield $907,500 or 90,750 shares. In return for
bringing Soriano and Malozemoff together, Lehman Brothers
were to have the right to acquire a 10 per cent interest in the results of Newmont's exploration work. Including the shares already owned, Newmont stood, as a result of this deal, to acquire
just under a 15 per cent interest in Atlas Consolidated, which
was the maximum ownership of a Philippine enterprise allowed
to a foreign corporation by Philippine law.

All this took time, and it was not until July, 1958, that the
Newmont geological study began. It was completed by the end
of June, 1961, at a cost to Newmont of $320,000. Involved in the
study were Newmont geologists P. C. Benedict and Jacques Claveau. Bob Searls took a great interest in the geophysical end of
the exploration, and a young Englishman, Alan Coope, now
manager of exploration in eastern Canada for Newmont, was responsible for the geochemical phase of the exploration.

Atlas exploration was the first use Newmont had made of
the techniques of geochemistry. Years later, geochemistry turned

out to be quite useful in exploration in Australia and South Africa, but at Atlas, it proved misleading. In the clay soil of the Lutopan area, geochemical samples yielded anomalies running into the thousands of parts per million of copper, but subsequent drilling showed only minor copper occurrences below these anomalies. The reason for the misleading results seems to have been that despite the heavy rainfall, the clay minerals fixed copper in some form that was not leached out and consequently built up to very high readings. In the end, the fact that the Newmont study located some 10 million tons of open pit ore was due as much to old-fashioned geology as to anything else.

In undertaking expansion, Atlas was able to finance capital expenditures of over $13 million. Production was increased, reaching an eventual total capacity of 21,000 tons per day in 1968. Profits also climbed, reaching an annual total of $4.5 million for 1963. Other exploration and improved mining methods made it possible to increase ore reserves to a grand total of about 147 million tons of ore assaying 0.74 per cent copper. Harry Nelems, a competent mining engineer, was brought in as manager, and with Newmont's assistance, a block caving technique was worked out, which made possible the economic underground production of ore from under the Lutopan pit.

In mid-1964 a typhoon dumped 36 inches of rain on the Atlas operations in a twenty-four-hour period. Much of a neighboring hill slid into the Lutopan pit, filling it to a depth of 100 feet with mud and water and putting it practically out of business. Open pit mining was gradually transferred to the Biga Road property where Newmont had found the 10 million tons of additional ore reserves.

Through at least 1966, the connection with Atlas Consolidated continued to be interesting and profitable for Newmont. From the exploration, right on through the mining, milling, and contract negotiations with the Japanese for sale of copper concentrates, Atlas at one time or another engaged the attention of nearly everyone on the Newmont staff. In particular, the Newmont people enjoyed the relationship with Colonel Soriano.

Andrés Soriano was a most unusual man as well as an outstanding hero of recent Philippine history. Already a prominent industrialist before the Second World War, Soriano refused to run when the Japanese invaded the Philippines, and chose instead to join the American Army in the fighting on the Bataan Peninsula. He left Bataan only when ordered to do so by General MacArthur, and he served with the general throughout the remainder of the war. Widely known for his courtesy, keen intelligence, and sense of fair play, his death in December, 1964, was a great loss to his company and to Philippine industry.

Beginning about 1966, the Philippine government began to place ever-increasing restrictions on Atlas Consolidated and other Philippine mining companies. These restrictions followed the same unhappy trend experienced in other countries whose principal source of income lies in natural resources. Artificial exchange rates, ever-increasing taxes, imposed wage settlements, difficulty of bringing in competent technical help, all these things and more made life increasingly difficult for Atlas. Furthermore, the rapport that Newmont had once enjoyed with Atlas seemed to have disappeared with Colonel Soriano. In brief, although Atlas continues to the present day to be an increasingly profitable copper producer, Newmont developed a dim view of the long-term outlook for this investment and liquidated its Atlas holdings.

Newmont's second foreign adventure of the mid-1950's brought the company back to Peru. This time, Newmont's participation in Southern Peru Copper Corporation did not grow out of a casual conversation, but rather out of a technical development in the field of geophysics that for some years gave Newmont a competitive edge over other companies in mineral exploration. This new method of geophysical exploration, called the induced polarization, or IP method, was developed under the direction of Dr. Arthur A. Brant of Newmont in the late 1940's and early 1950's at a laboratory at Jerome, Arizona. Development of Newmont's IP technique is described in Chapter 14 on exploration, and it is enough to say here that the IP method showed promise of being able to delineate porphyry copper orebodies with a degree of probability not heretofore attainable except by actual drilling of the orebody.

Although Newmont tried hard to keep the discovery of the IP technique a secret, news of it inevitably leaked out, and one of those who heard about it was Herbert Kursell, a senior mining engineer for American Smelting and Refining Company. Actually, at the time he heard of the technique, Kursell had retired from Asarco and was engaged from time to time as a consultant for the Cerro Corporation, then called Cerro de Pasco Copper Corporation. Years previously, Kursell had been closely involved in the exploration that developed the Toquepala porphyry copper orebody of Asarco in southern Peru. Kursell was later involved in a reappraisal of Cerro's porphyry orebody, called Cuajone, located not far from Toquepala.

Having been a close friend and associate of Malozemoff's father many years previously, Kursell called on Malozemoff to say that if what he had heard about Newmont's IP method was true, it might be extremely useful for delineation of the Cuajone orebody. At the time, the Cerro Corporation was somewhat short of cash, and to carry on the large-scale drilling campaign that would be required for Cuajone could have become quite burdensome for Cerro. Malozemoff brought the idea to the attention of Fred Searls, and in due course, Searls, Malozemoff, and Kursell got together with Robert P. Koenig, president of Cerro, to discuss the matter. Among the four of them, a deal was worked out whereby Newmont would furnish up to $1.5 million for exploration of Cuajone and would also contribute the new geophysical technique as a part of the job. Newmont's first survey of Cuajone was directed by Larry Baldwin, a most competent young geophysicist who tragically was killed in the crash of a helicopter in Alaska several years later. In return for the cash contribution and the use of the IP method, Newmont was to receive a 15 per cent undivided interest in Cuajone.

With Newmont paying the bills, mobilization for a drilling campaign proceeded, but by the time that nearly a million dollars had been spent, there was still very little drilling done. An intensive discussion period followed, after which Newmont agreed to put up another $1 million in return for an additional 10 per cent of the venture that might develop out of the exploration. Shortly thereafter, the drilling and geophysical exploration at Cuajone was completed and had outlined around 600 million tons of ore running about one per cent copper.

At about that time, in 1954, Newmont and Cerro heard that the Export-Import Bank of Washington had made a commitment to American Smelting and Refining Company for a loan of $100 million for the purpose of equipping the Toquepala orebody in Peru for mining, milling, and smelting, a project that required an estimated $200 million or more to carry out. Thus encouraged, Malozemoff and Koenig promptly submitted a similar application to the bank for the Cuajone project.

While the Export-Import Bank was considering this application, the Asarco management and its board came to the conclusion that the investment that Asarco was in a position to commit, plus the Export-Import Bank loan, was not sufficient to cover the entire cost of the Toquepala project. Deciding that it might be well to bring in a partner, the Asarco management concluded that the Phelps Dodge Corporation would be the most likely prospective partner.

It happened that at that very time, R. P. Koenig had invited Messrs. Cates and Page of Phelps Dodge to have lunch with him for the purpose of discussing the possible participation of Phelps Dodge in developing the Cuajone project. Between the issuance of Koenig's invitation and the lunch itself, the Ex-Im Bank announced publicly the agreement to loan $100 million to Asarco for the Toquepala project. Before sitting down to lunch, Cates told Koenig that he felt he should mention that on that very morning Asarco had called him to ask if Phelps Dodge would be interested in considering joining in the Toquepala project. Cates went on to say that because he had already done some thinking about developments in Peru, he was able to tell Asarco at once that Phelps Dodge would be unwilling to consider participating in either Toquepala or Cuajone unless the three orebodies, that is, Toquepala, Cuajone and Quellaveco, could be developed in one project under one ownership. That ended the discussion on mining projects, and during the remainder of the luncheon Koenig reports, "We discussed baseball and taxes."

About 3:00 that afternoon, Kenneth Brownell of Asarco called Koenig to mention that he had again talked with Cates and Page and that they wanted to get together to discuss reaching a policy decision on the consolidation of the development of the three orebodies. Such a meeting was held a day or so later among Messrs. Brownell, Cates, Koenig and Malozemoff, and agreement in principle was quickly reached. The proposition was then turned over to the staffs of the four companies, and in what one participant later referred to as "the battle of the burning slide rules," the project took shape under the name of Southern Peru Copper Corporation.

By agreement, American Smelting and Refining would contribute the Toquepala and the Quellaveco orebodies, which together were about the same size as Cuajone. In turn, Cerro and Newmont would contribute the Cuajone orebody, which was then owned 75 per cent by Cerro and 25 per cent by Newmont. Phelps Dodge's contribution was to be the provision of a larger share of cash than the other companies would contribute, plus Phelps Dodge's expertise in open-pit mining. As an inducement to Phelps Dodge to take this action, it was agreed that when Southern Peru began to pay off, Phelps Dodge would recover part of its investment in advance of the other companies. Eventually, after a minor realignment, Southern Peru's ownership settled as follows:

American Smelting	$51\frac{1}{2}\%$
Cerro Corporation	$22\frac{1}{4}\%$
Phelps Dodge	$16\ \ \%$
Newmont	$10\frac{1}{4}\%$

This ownership of Southern Peru gave the four companies possession of three copper orebodies containing something over one billion tons of ore assaying around 1 per cent copper.

Before Southern Peru really got rolling, however, Louis Cates insisted on obtaining check samples on at least the Toquepala orebody, plus an analysis of the conclusions reached on the basis of the drilling and geophysical work already done. Cates further requested that the samples should be obtained through a system of drifts and raises, because only then would he feel assured of the value of the orebody. To direct this program, Newmont called on Henry Krumb once more. Still quite active, Krumb, then seventy-seven years old, could not resist the challenge and returned once more to the battle. As it had always been, Krumb's advice was most helpful. For example, he refused to be dismayed if samples from a raise under one drill hole were discrepant. He would wait for results from eight or ten raises before reaching any conclusion, and in general, the conclusions reached corroborated the earlier exploration work.

Southern Peru turned out to be Henry Krumb's last assignment for Newmont. The old gentleman died on December 27, 1958. One of the most respected and honored members of the mining industry, Krumb had to his credit an extraordinary number of accurate evaluations of mineral properties, especially in porphyry coppers. Typically, Krumb left more than his memory to the mining industry, for in a bequest that eventually came to $18 million, he endowed and equipped completely new facilities for the College of Mines of Columbia University, where he himself had received his education. It is entirely fitting that Henry Krumb's name should thus be associated with the latest and one of the largest of the world's porphyry copper properties.

One of the first problems of the Southern Peru project was the selection of a manager. Following considerable discussion among the four partners in the enterprise, agreement was reached on the selection of Edward McL. Tittmann, who had achieved an excellent reputation in managing Asarco's western smelters. Tittmann directed the enterprise during the first critical years of construction and production, and Malozemoff credits him with the fact that the venture began with unusual smoothness for so large a project in a foreign country.

Technically and economically, Southern Peru has proven to be extraordinarily successful. Newmont's investment in Southern Peru of $13.2 million had all been repaid by 1966, and as of the end of 1971, Newmont had received in addition a total of $26,866,000 in dividends. In 1971, despite low copper prices and numerous work stoppages, Southern Peru showed a net income of $26,214,189, out of which Newmont received $3,047,000 in dividends.

Southern Peru now faces a political problem that is by no means as susceptible of solution as are technical or economic problems. In 1968, the elected government of Peru was displaced by the military in a coup that established a junta of generals in command of Peru's Federal government. In what appears to be a sincere and determined effort to improve the lot of the majority of Peru's citizens, the junta has established a series of new social directives that have grown steadily more far-reaching in their economic consequences to the country.

Under current Peruvian law, established by the junta, anyone owning a mineral concession in Peru must present a specific plan for its development within a certain time or lose the concession. Pursuant to the terms of the law, Southern Peru Copper Corporation lost to the Peruvian government in 1971 its ownership of the Quellaveco orebody. However, earlier Southern Peru reached an agreement with the junta in December, 1969, on a specific plan for the development and equipping of the Cuajone orebody. Total cost of this project, which would include an electrolytic refinery in Peru, is expected to reach around $500 million. By agreement in June, 1971, Southern Peru was to have spent, or committed, for the Cuajone project an additional $48 million by the end of 1972, and this work is proceeding. Negotiations were still in progress in early 1972 in an effort to arrange financing for the entire project.

Palabora Mining Company Limited, the third of Newmont's major foreign involvements during the mid-1950's, is an interesting study, deserving of much greater space than can be given it in this chronicle. As copper orebodies go, the Palabora deposit is certainly unusual, if not unique, and to solve its problems of exploration, ownership, technology, and financing called upon all the resources that Newmont and its associate could bring to bear. Palabora has been, and is, a resounding success, and it has yet to experience any political interference.

Signpost for the Palabora orebody was a rocky knob projecting above the flat-lying plain just west of the boundary of the famous Kruger National Park game sanctuary in the northeastern Transvaal, the Republic of South Africa.

Known locally as the Loolekop, this prominent outcrop has attracted interest on the part of mineral explorers for centuries past. Actually, the Loolekop lies at the approximate center of what is now recognized as a mineral complex covering an area roughly 5 miles long and 2 miles wide. Located near the tiny town of Phalaborwa, about a mile west of the entrance to the Kruger National Park, the Palabora complex presents a fascinating picture to the geologists who have examined it.

At the center of the complex is an intrusive composed of a limestone that is of magmatic origin, an odd circumstance for anyone accustomed to thinking of limestone as only of sedimentary origin. Known as carbonatite, this magnesian-limestone carries a wide variety of minerals including magnetite, apatite, biotite, chondrodite, phlogopite, and various rare minerals. The carbonatite also carries several copper minerals, the most abundant of which are chalcopyrite and bornite, plus small amounts of a copper-iron sulfide called valeriite. The latter is a soft, greasy mineral, usually intergrown with magnetite, and is represented roughly by the formula $Cu_3Fe_4S_7$.

The carbonatite directly under the Loolekop assays about 1.26 per cent copper. However, there is no resemblance between the Palabora ore and the Arizona porphyry coppers. For example, the copper in the carbonatite central zone is not truly disseminated. There is some fine-grained chalcopyrite, but most of the copper occurs in the form of blebs and veinlets of chalcopyrite and bornite, a circumstance that made the Palabora orebody extremely difficult to sample accurately. There were instances of drill cores yielding copper assays as high as 4 per cent, to be followed immediately by cores assaying less than 1 per cent copper. It was only by discounting the high-grade assays that an accurate picture of the Palabora orebody could be put together.

Surrounding the central carbonatite is an elliptical low-grade copper zone about 450 feet wide and 1,700 feet long, carrying about 0.73 per cent copper. This zone consists mainly of phoscorite, which is a combination of olivine, magnetite and apatite mined locally for its apatite content by Phosphate Development Corporation Limited, called FOSKOR for short, which supplies South African producers of fertilizer.

Surrounding the second zone is a third zone consisting mostly of micaceous pyroxenite which carries concentrations of vermiculite. This zone does contain copper, but is still lower grade than preceding zones.

The final zone of the Palabora complex consists of syenitic rocks, largely feldspathic pyroxenites, still containing some copper. The entire complex is set in the regional Archaean granites.

Although there has been some mining and smelting activity at or near the Loolekop for possibly a thousand years, the first really scientific approach to the area was made by the famous Dr. Hans Merensky in 1912. Merensky is remembered best as the discoverer of the diamond field in Namaqualand near the present O'okiep mining project, and, later, of the "Merensky Reef," now famous for its platinum mines.

Merensky discovered that the Loolekop area contained high percentages of phosphate and also of vermiculite, both of which had an immediate commercial use in South Africa. Merensky also recognized the presence of copper, but it appeared to him at that time as too low in grade to be at all interesting.

Merensky did form a company to produce rock phosphate

from the Loolekop area, but not until the outbreak of World War II, which cut off shipments of Moroccan phosphates to South Africa, did his company, South African Phosphate Company, find ready markets in the country.

In 1951, the South African government had set up FOSKOR to supply South African fertilizer producers. As mining by FOSKOR proceeded, it became apparent that there were available in the Loolekop area enough supplies of phosphate to meet any projected fertilizer requirements in the republic.

In addition, a company called Transvaal Ore Company was set up by the Hans Merensky Trust to mine and to produce vermiculite in a part of the Loolekop area that included the central core of the ore deposit itself. Even at this late date, the presence of copper in the ground that was being mined for phosphate and for vermiculite failed to arouse any interest.

Anyone active in the mining industry in the late forties or early fifties will remember the tremendous drive conducted by the Atomic Energy Commission to locate additional supplies of uranium. Every mining company, every museum, in fact, anyone with a collection of rocks anywhere was asked to run a Geiger counter over his samples or specimens to see whether or not some unsuspected occurrence of uranium might exist. This was why a unit of South Africa's Atomic Energy Board found itself in the Transvaal Museum in Pretoria running Geiger counters over the mineral specimens on display there. Their boredom with this unexciting assignment disappeared when one rock caused the Geiger counter to give off a crescendo of clicks. The label on the specimen gave its source as the Phalabowra District.

In February, 1953, the South African government set up an exploration program to track down the source of this specimen. A scintillometer survey was flown over the Phalabowra area, and over a period of three years the South African exploration teams drilled seven holes into the Loolekop and also drove some 3,000 feet of drifts into the knob.

Although a certain amount of radioactivity was encountered, either over the Loolekop or in the samples taken from under it, there did not appear to be enough radioactive material present to make its exploitation worthwhile. The exploration did, however, encounter what appeared to be at least marginal grades of copper.

At about this time a Newmont affiliate and an English mining company set out on exploration programs in South Africa that brought the two companies into direct confrontation. In 1953, the Rio Tinto Company, Limited, of England, long active in the pyrite mines in Spain, obtained permission of the South African government to explore in the neighborhood of Phalabowra. The geophysical part of the exploration was directed by Oscar Weiss, who had become internationally famous because of the magnetic and gravimetric geophysical work he did in South Africa that led to the discovery of the gold mines in the Orange Free State. The work done by Weiss clearly indicated the presence of commercial quantities of copper, and accordingly Rio Tinto moved to tie up claims in the Loolekop area. For this, they approached a Mr. Von Bülow, the head of Hans Merensky Trust, which at that time was operating the vermiculite property of Transvaal Ore Company.

Also in 1953, Henry DeWitt Smith in London heard from a friend of his that a government exploration program in South Africa had turned up copper near Phalabowra. O'okiep Copper Company Limited and Tsumeb Corporation Limited already had in being a jointly owned exploration arm called Safari Exploration Limited, and Henry Smith therefore asked M. D. Banghart and P. C. Benedict, on behalf of Safari Exploration, to go to Phalabowra and examine the area. Benedict later wrote in his report: "Every effort should be made to acquire this property," and Safari did, indeed, make every effort, but to no avail.

On the strength of Benedict's report, Safari, acting through O'okiep, approached the South African government for a mining lease subject to carrying out further exploration leading eventually to production. By coincidence, Oscar Weiss, only a few days before Safari made its request, had also approached the government for the same purpose, after Rio Tinto obtained a verbal option from Von Bülow on the claims that were held by the Hans Merensky Trust. Under this, Rio Tinto had agreed to pay the trust £500,000 for its claims over the central part of the Loolekop. Von Bülow felt that he was committed by this verbal agreement, and when Benedict attempted to acquire the same claims, Von Bülow told him he felt obligated to Rio Tinto, although no option document had as yet been signed.

However, Safari did apply to the South African government for a mining license just as Rio Tinto had done. This called for careful consideration on the government's part. While the matter was under consideration, Val Duncan, the chairman of Rio Tinto, visited Malozemoff (as representing Newmont, not Safari) with the suggestion that it would seem sensible for the two companies to combine their efforts and not attempt to obtain exclusive leases. Malozemoff was interested, so he and Duncan discussed various aspects of the situation at some length. The discussions finally broke down on the question of which of the participants to a joint venture might wield the management responsibility.

Some months later, still before the South African government had reached a decision, talks between Newmont and Rio Tinto were resumed with the suggestion by Malozemoff that a type of management, independent of either of the owners of the project, might well be established. This would somewhat resemble the management philosophy of Southern Peru Copper Corporation, in which Newmont was participating. However, Val Duncan voiced the opinion that this would be "a kind of Soviet," and once more the discussions broke down.

Shortly thereafter Rio Tinto and Newmont received word that the South African government was quite unhappy at the impasse. The government recommended that the two companies get together on a joint venture of some kind because the government was unwilling to divide the property in any way.

On receiving this news, Malozemoff and Roy Bonebrake, Newmont's chief counsel, went to London in June, 1956, with the firm intention of reaching an agreement. This was done within two days, on the basis of a quite complex but nonetheless workable financing formula. In the end, Rio Tinto was given the management responsibility of the new company that was to be formed, in part because of its uncompromising insistence on this point, and because Rio Tinto did, after all, hold the trump card of the critical claims of Transvaal Ore Company in the heart of the Loolekop area.

Once the major elements of the deal were worked out, an organization called Palabora Mining Company Limited was set up to drill the Palabora property thoroughly and to do preliminary metallurgical test work. To supervise this work, Newmont provided the services of Edwin S. W. Hunt, who had been a mining engineering consultant for Newmont for some time, and was most recently assistant general manager at O'okiep. Under Hunt's direction, there were drilled 111 inclined diamond drill holes on 250-foot squares over the Palabora ore zone. Total footage drilled was 135,937 feet. Core recovery came to 96 per cent, which was unusually high for that kind of campaign.

To get a more or less representative sample for pilot plant testing, the company drove a tunnel into the hillside at the 400-foot level.

This drilling campaign indicated an ore reserve from the surface down to 1,000 feet of 315 million tons averaging 0.68 per cent copper. The indicated waste to ore ratio was 0.89 to 1. Also in these reserves, there were approximately 80 million tons of iron recoverable as magnetite.

Fairly early in the drilling campaign, Malozemoff and Ed Hunt became aware of still another peculiar element in the Palabora orebody. There is no sharp cut-off to the ore zone. In other words, the content of copper simply diminishes as one goes outward in any direction horizontally from the central core. This indicated that one could go very deep with an open pit, because the walls of the pit could simply be extended into ore of continually lower grade. No material that would have to be mined would be completely barren. Therefore, some of what would ordinarily be disposed of as waste contained enough copper to repay milling costs, thus reducing the tonnage of waste to be discarded. Because of this circumstance, both Malozemoff and Hunt were quite enthusiastic over the possibilities at Palabora.

But not all members of the Newmont management, or, apparently, of the Rio Tinto management, shared this enthusiasm. The whole enterprise presented a number of geological and economic problems. The ore grade was low by any standard, metallurgy presented serious problems, and finally the formidable nature of financing the project was just beginning to be fully appreciated. For example, at one point, difficulty in concentrating the Palabora ore endangered the whole enterprise. The initial flotation tests that had been done on drill cores indicated erratic results. Tests would sometimes give acceptable results, and then cores that looked and assayed the same would yield unacceptable results. The problem was finally solved when Malozemoff had Han Hartjens, an employee of Newmont's research laboratory at Danbury, Connecticut, and formerly of American Cyanamid Company, run flotation tests, using a standard procedure, on a sample taken from every 10 feet of core in all the drill cores from the Palabora orebody.

This standard procedure resulted in supplying a means of identifying the refractory ores by plotting recovery of copper against the source of each sample. It was possible to delineate a sausage-shaped zone in the central core of the orebody where the standard flotation procedure gave low recovery. Furthermore, it was possibly to identify the source of the trouble as the mineral valeriite, which simply did not respond to flotation.

This indication of low recoveries in the center portion of the orebody caused considerable consternation among some of the English and American board members of Palabora, who did not appreciate at first the full significance of the test work. Actually, Hartjens's work provided a kind of road map to the nature of the orebody, and at least the problem could then be identified, and recovery of valeriite studied separately.

One by one, all objections were overcome and the decision was made to go ahead. At Newmont's insistence, the engineering work was to be done by contract in the United States. Ed Hunt

and others from Newmont studied the potentiality of various
contractors before any agreement was reached. In the end, a
combination was worked out as a joint venture between the Bechtel Corporation and Western Knapp Engineering Company.
The basis for this was a desire to use the engineering ability of
the Bechtel Corporation combined with the mining and milling
experience and abilities of Western Knapp. Frank Buchella, a
man with great experience in copper smelting and, at that time,
the general manager at San Manuel for Magma Copper Company, was selected to supervise the engineering work along with
Ed Hunt. These two, together with men from Bechtel and Western Knapp, went to Palabora and in about a week had laid out
the location of the various buildings and had agreed upon the
scope of the work. There has been no major deviation from the
plan these men originally laid out, which testifies to the quality
of the job they did.

Based on the preliminary engineering work, it was estimated the total cost of equipping the property, including housing, would come to about $103 million. An attempt to raise
funds ran at once into another snag. Newmont, as the American
representative in the enterprise, agreed to approach the Export-
Import Bank or the World Bank for part of the financing required for Palabora. Both banks turned out to be extremely cool
toward the project, once they learned its location. The unofficial
but nonetheless firm attitude of the United States government
and of the World Bank seemed to be one of disinterest in helping further an enterprise in the Republic of South Africa. In the
end, financing was obtained elsewhere, mainly through the skill
and resourcefulness of Val Duncan and Roy Wright of Rio
Tinto.

The money was actually obtained as follows:

Common share capital subscribed	$ 40,779,200
1st Mortgage debentures, sold in Africa	17,675,000
West German loan	26,950,000
Equipment loans and short-term bank loans	17,789,800
Total	$103,194,000

Of the common share capital, $34,654,000 was subscribed in purchase of shares by several subsidiaries of Rio Tinto, by Newmont, by American Metal Climax, Inc., Union Corporation, Selection Trust Limited, and Seltrust Investments Limited, the last three being South African companies. The last four companies were brought in as shareholders of Tsumeb and O'okiep, and in turn, as shareholders of Safari Exploration.

The first mortgage debentures, bearing 6.5 per cent interest, with stock attached, were taken up by various banks in Africa and Europe. The German loan came from the Kreditanstalt für Wiederaufbau. Referred to as KFW, this agency, as the name suggests, is a kind of West German Reconstruction Finance Corporation. Of the loan from KFW, the West German government guaranteed 90 per cent against political risks, and 80 per cent of the loan against economic risk. The sponsoring shareholder guaranteed the remainder of the loan, which bore interest at 6 per cent. The loan was further secured by second mortgages on the property and by a pledge of all the copper to be delivered under a copper sales contract made with Nordeutsche Affinerie, of Hamburg, Germany. This German smelter agreed with Palabora to take 36,000 tons of blister copper annually for the first five years of production, then 30,000 annually for the next fifteen years. In the beginning it was Palabora's intention to sell the remainder of the copper it would produce on world markets, but before long a decision was made to refine at Palabora the copper not sold to Nordeutsche, and produce the electrolytic copper in semi-fabricated form, partly as wire bar and partly as continuously cast rod, to be sold mainly in South Africa.

During the course of the negotiations that led up to actual operations at Palabora, the Rio Tinto company was merged with Consolidated Zinc Corporation, an Australian company. This merger resulted in an extraordinarily large complex of companies, the parent company becoming known as Rio Tinto-Zinc Corporation, or RTZ for short. In the British fashion, there were established thereafter a long string of subsidiary com-

panies operating in various countries and known as the RTZ
Group.

The financial structure of the Palabora project seems rather
complex to those of us more familiar with American mining or-
ganizations. Majority ownership of Palabora Mining Company
Limited is in the hands of a separate company called Palabora
Holdings Limited, which in turn is owned by Rio Tinto-Zinc
and Newmont. Participation in Palabora Mining Company is as
follows: Rio Tinto Mining Company of South Africa owns 3.48
per cent, Newmont owns 2.66 per cent, the general public and
certain others own 32.45 per cent, and Palabora Holdings Limited owns 61.41 per cent.

Palabora Holdings Limited has only the function of paying
out money and of taking in dividends and distributing them as
indicated by ownership. Palabora Holdings has no assets other
than the shares of Palabora Mining Company that it owns, plus
varying amounts of cash, nor did it have liabilities or commit-
ments other than the repayment of a loan of $1,160,000 made in
the early days of the enterprise by Newmont. This loan was repaid partly in cash and partly in the form of stock in Palabora
Holdings Limited. After this loan had been repaid and other
changes in holdings made, ownership of Palabora Holdings became and remains as follows: the Rio Tinto group, 57.73 per
cent, and Newmont, 42.27 per cent.

Beginning production in 1966, Palabora has steadily increased not only its output but the range of products turned out
from the property. The following table shows the production of
various items for 1971.

PALABORA PRODUCTION, 1971

	Metric Tons
Cathode copper	60,703
Wirebars	23,543
Coiled rod	36,046
Magnetite concentrate	951,531
Vermiculite	132,071
Sulfuric acid	74,991

In addition to all the foregoing, research is now proceeding looking toward production of uranium oxide.

It is difficult to avoid overenthusiasm in describing the accomplishments at Palabora. One of the lowest-cost producers of copper in the world, Palabora has become a well-knit, highly polished operation that will stand comparison with any large copper mining, smelting, and refining operation anywhere. Palabora's net income for 1970 was $46,941,000, and for 1971, $32,357,000, of which Newmont received $7,335,000 in dividends. Alone among the three ventures abroad described in this chapter, Palabora faces a future unclouded by anything other than the normal vagaries of the copper market.

Nevertheless, since the three enterprises were initiated, Newmont became more and more wary of foreign involvement, finally deciding in the early 1960's to increase the proportion of its income derived from North America. The story of how this shift was made from major emphasis on foreign income back to North American sources is one of the liveliest and most interesting segments of the Newmont history.

Magma's new refinery

13 Financing Expansion at Home

In the catalog of Newmont ventures and investments, the decade of the 1950's is pre-eminent from the point of view of sheer number of major Newmont involvements. However, in terms of consolidation of assets and of company growth, the record of the ten years beginning about 1960 far surpasses any similar period in the whole Newmont history.

Of the many problems facing Newmont in the early 1960's, two were outstanding. First, there was the question of how to build up the North American sources of income for Newmont, an imperative then clearly recognized by almost everyone in the company and supported by the Newmont board of directors. Second, there was the problem of financing what seemed the astronomical costs of the new kind of major mining enterprise that Newmont now felt itself inclined to tackle. By a most ingenious but entirely logical procedure that is well worth examining in some detail, Newmont was able to solve both problems in closely related, rather than separate, maneuvers.

Only ten years previously, in 1950, many in the Newmont management and its board would have been horrified, and to some extent were, at the thought of Newmont or one of its subsidiaries risking as much as $75 or $100 million on a single mining venture, San Manuel, for example. In ten short years, this attitude had largely disappeared, in part because of Plato Malozemoff's willingness to think in larger terms, in part because of the new spirit at work in the board, and in part because circumstances forced the acceptance of ever larger capital expenditures. Only twenty-five years ago, a man could still get a grubstake from the owner of the local general store and go out on his own and locate a mineral property that could make him rich. Charlie Steen and Vernon Pick did just that in two separate uranium mining enterprises.

Now, however, a grubstake from the grocer just isn't enough. Nowadays, it takes a friend at Chase, or better yet, friends at half a dozen major banks and insurance companies to supply the millions of dollars that are needed for development of a mineral property that has a real chance of being profitable.

Newmont had such friends, in commercial banks, in investment banking houses, and in the insurance companies, and they were willing to back Newmont's judgment with cash. Friendship may in the past have caused a Bernard Baruch to offer a million dollars to develop a mine such as the Getchell, but today friendship serves as little more than a means of introduction to the right people. The hundreds of millions of dollars required for today's mining ventures are forthcoming only to those companies in whose judgment and credit standing the lenders have the utmost confidence.

The confidence of the investment community in Newmont was displayed to good advantage in the series of moves by which Newmont acquired Magma Copper Company and by which the tremendous current expansion program at San Manuel and Superior has been financed. In 1961, although Newmont owned 21.9 per cent of Magma Copper Company's shares, the investment returned nothing to Newmont, inasmuch as under the

terms of the R.F.C. loan to Magma, the company paid no dividends. For example, Magma showed a profit of $9.5 million for
1961, but surplus funds that might have been used for dividends
had instead to be applied to repayment of the R.F.C. loan, of
which $65 million was still outstanding.

As the result of a suggestion by André Meyer, the principal
partner of Lazard Frères and also a director of Newmont, Malozemoff became aware of certain possibilities. First, Magma, both
at Superior and at San Manuel, looked like one of the greatest
potential assets in the Newmont portfolio. Second, if Magma
could resume paying dividends, and if Newmont could increase
its ownership of Magma, it would greatly increase the proportion
of Newmont's income from North American sources. Third, if
Newmont owned a greater share of Magma, Newmont's excellent credit standing might enable Magma to refinance its loan
on better terms. Plans were accordingly developed to acquire up
to about 80 per cent interest in Magma to enable consolidation
with Newmont's accounts for tax purposes.

Not all of Malozemoff's associates were enthusiastic about
the idea. Franz Schneider did not like it because he foresaw that
such a move might involve both companies in difficulties with
the Department of Justice, as it eventually did. Furthermore,
Magma had always operated quite independently of Newmont,
and some of the Magma staff preferred to keep it that way. On
the other hand, André Meyer strongly favored a full merger
with Magma, which Malozemoff was reluctant to attempt at the
time because it might have resulted in forcing some minority
shareholders to accept a share exchange that they did not regard
favorably.

In the end, Malozemoff, working with André Meyer, Roy
Bonebrake, and others, structured a proposition for acquiring
approximately 80 per cent of the Magma shares, which was presented to the Magma shareholders in April, 1962. At that time,
Newmont offered to Magma's shareholders, other than Newmont, three quarters of a share of a new issue of Newmont convertible 4 per cent cumulative preferred stock in exchange for

each Magma share. The offer was good until 809,794 shares of Magma had been submitted in exchange for the Newmont preferred. The latter was convertible into Newmont common shares at a price of $90 per common share if converted on or before April 30, 1967, and at $100 per share thereafter, through April 30, 1977. The offer was well received, and when the transaction was completed, Newmont had become the holder of 80.6 per cent of the outstanding Magma common stock.

With the help of Lazard Frères, Roy Bonebrake set about attempting to refinance the R.F.C. loan, which by May of 1962 had been reduced to $54 million. Following a not too difficult series of negotiations, Magma arranged with The Prudential Insurance Company of America a loan of $54 million at interest of 5½ per cent and maturity extended to 1982, well beyond the 1973 maturity date of the existing R.F.C. loan. Terms of the Prudential loan permitted payment of cash dividends by Magma and the use of surplus funds for further development work at the property. Accordingly, in February, 1963, the R.F.C. loan was paid off entirely out of proceeds of the Prudential loan, and in the same year Magma declared its first cash dividends since 1959, amounting to $0.85 per share for the year, or a total of $3,556,-222. Magma also split its stock in 1963 in the ratio of 3 for 1. Thus Newmont substantially increased its domestic earning capacity, and much more was to come.

To understand the next major development at the San Manuel properties of Magma, one must form a picture of the San Manuel porphyry copper orebody. To put it in the simplest terms, the San Manuel orebody somewhat resembles the lower half of an inclined thick-walled pipe, the upper half of which has been sawed off. The result is a sort of trough trending roughly northeast-southwest and dipping about 20 degrees to the southwest. What would have been the upper half of the orebody to complete the resemblance to a pipe was cut off by a large fault, the San Manuel Fault, dipping at a rather flat angle to the southeast. Movement on this fault, whether up or down, had been apparently very large, possibly thousands of feet. In the drilling campaign to delineate the San Manuel orebody, there was little or no effort made to locate the missing upper half, because there was little indication that it still existed.

The possibility that there just might have been an upper section of the San Manuel orebody that had not been destroyed by erosion intrigued a consulting geologist in Tucson named J. David Lowell. Studying all available information, Lowell came to the conclusion that a search for additional ore was worth a try on the assumption that the upper half of the orebody had been shifted downward and to the west. If Lowell's interpretation were correct, the missing orebody would be found under certain claims west of the San Manuel property 7 miles east of Oracle. Lowell took his ideas to Corbin Robertson, manager of the extensive oil interests of the Cullen family of Houston. Becoming interested, Robertson organized a company, Quintana Minerals, Inc., and Lowell went to work in August, 1965, drilling at the point under which he believed the missing orebody to exist.

Whether by good luck, by geological skill, or by both, the first hole struck ore-grade mineralization at a depth of 2,500 feet. Twenty-one additional holes were drilled, and as Lowell himself put it, "The unique opportunity for extrapolating geology and symmetry of mineral zones from the San Manuel deposit has made it possible to intersect the Kalamazoo (Robertson's name for the new orebody) with every hole completed to date." One hundred per cent effectiveness in a drilling campaign is, as far as is known, unparalleled in the mining industry.

Naturally, Magma heard about this discovery, and obviously the best way of mining the Kalamazoo orebody was from existing San Manuel workings. However, Robertson was not interested in a deal involving Magma stock, and Magma, alone, did not have sufficient cash to meet Robertson's terms, and thought his price too high in the first place. Thereupon, Newmont entered the discussion, and following long and difficult negotiations, largely between Malozemoff and Robertson, Magma purchased the Kalamazoo orebody from Quintana Minerals, for $27 million in cash and stock equivalent, including both Newmont and Magma shares. Thus, by acquiring over 560 million additional tons of ore assaying 0.72 per cent copper, Magma had assured itself of a much longer life and an opportunity for expansion.

This doubling of ore reserves for San Manuel did, indeed, put quite a different face on the future. After considerable study, the Magma board in July, 1968, authorized expansion of facilities at San Manuel to mine and mill 60,000 tons of ore per day, an increase from the original 30,000 tons and a subsequent increase to 40,000 tons per day. Furthermore, the board authorized the expansion of mining facilities and milling capacity at Superior (as described in an earlier chapter), and Magma, in addition, launched an even greater expansion in the San Manuel smelter than the expansion decided upon in 1968. (The San Manuel smelter now has a capacity of 1 million tons of concentrates annually, substantially more than the Magma and San Manuel mines can produce.) All of the foregoing required an investment of approximately $250 to $275 million.

Financing so large an expenditure was not an easy matter, although by the end of 1967, Magma had reduced the Prudential loan from its original $54 million to $21 million and had also paid off the San Manuel townsite mortgage debt of about $9 million. However, at that time, Magma had borrowed another $15 million from various banks as part payment for the Kalamazoo orebody (repaid by March, 1970). Malozemoff and André Meyer believed that Magma's financing problems could be handled more effectively if the company were merged entirely into Newmont. Accordingly, in March, 1968, Newmont exchanged with the remaining Magma shareholders, other than Newmont itself, 0.85 share of a new issue of $4.50 cumulative convertible preferred stock of Newmont for each share of Magma stock held. Each share of the new preferred stock was to be convertible into 1.25 shares of Newmont common stock at any time. However, the company has the right to redeem the preferred stock at any time after January 1, 1974, at a redemption price of $104.50 during 1974, and decreasing by $0.50 on each January 1 thereafter until January 1, 1983, at which time and thereafter the price of redemption will be $100.

This transaction was duly completed, and the reorganized Magma Copper Company is now a wholly owned subsidiary of Newmont, although Magma's operations continue to be managed in much the same way as formerly. However, as of January 1, 1972, Wayne H. Burt became president and chief executive officer of Magma, replacing Wesley Goss, who became chairman of the board. Goss replaced Roy Bonebrake, who retired on December 31, 1971. Wayne Burt joined Magma in 1969 as assistant general manager and was promoted to executive vice president of Magma in March, 1971. Burt came to Magma from the Utah Division of Kennecott Copper Corporation, where he was general superintendent of smelting and refining. He is also a member of the Magma board. His accession to leadership coincides with a dramatic broadening of the scope of Magma's operations.

Early in 1969, it occurred to Malozemoff that with the expanded production possibilities of Magma, there existed a further opportunity to increase Magma's earning power. When the expansion at Superior and San Manuel is completed in late 1973, the company will have the potential of producing around 200,-000 tons of copper annually. Insofar as the copper industry is concerned, this puts Magma "in the big time," as one member of the Magma staff expressed it.

At Malozemoff's request, there was instituted a market study of copper consumption and fabrication patterns in the United States. The study also included an examination of Magma's past refining and selling practices. The study revealed that the center of gravity of the copper-consuming industry of the United States was moving steadily westward, so rapidly in fact, that markets for semifabricated copper were developing in the Middle West and Far West at points that could be conveniently reached directly from San Manuel.

Traditionally, Magma's blister copper had been refined elsewhere by another company and had been sold largely along the eastern seaboard. Investigation showed that freight savings could be made and refining costs reduced if Magma were to refine and sell its own copper.

As a result, the Magma board authorized in mid-1969 the construction of a copper refinery and a continuous-cast rod mill at San Manuel at a cost of approximately $31 million, not including the cost of necessary additional housing at San Manuel. The refinery was designed and built for a capacity of 200,000 tons of electrolytic copper annually. The design also provides for expansion to perhaps 300,000 tons of copper annually if outside supply of concentrates justifies that expansion. The Bechtel Corporation of San Francisco handled design and construction of the refinery.

To achieve the most modern design possible in an electrolytic refinery, Newmont and Magma engineers, together with consulting experts, visited most of the world's copper refineries, principally in Africa and in Japan, as well as in the United States and Canada. The new refinery entered production in December, 1971.

In keeping with the information developed during the market studies, the refinery will not produce wire bars, but instead produces cathode copper for sale, and by means of a procedure developed by the Southwire Company of Carrolton, Georgia, about half of the refinery output is in the form of continuous cast rod. This rod, which is also being produced in a number of installations including Palabora in South Africa, will be in the form of 5,000- to 16,000-pound coils of $5/16$-inch rod suitable for use by manufacturers of copper cable. Provision of rod in long coils enables the cable manufacturers to turn out a superior product more efficiently.

A demand for this continuously cast rod already exists in the Southwest and the Middle West. Certain long-term contracts for sale of Magma's rod were negotiated before the refinery had been completed. Because of its extra production cost and its attractiveness to cable manufacturers, continuous-cast rod brings a premium over the copper wire bar price.

On the heels of the decision to provide a refinery at San

Manuel, there began to develop, not only for Magma Copper Company, but for the entire U. S. copper mining industry, a situation without precedent in all the industrial history of the United States. To summarize, Magma and the other U. S. copper mining and smelting companies were put under tremendous pressure by Federal and State action to undertake mammoth expenditures for pollution control that would be uneconomic in nature and that are as yet unjustified by any demonstrated requirement for protection of public health and welfare.

Granted that the copper smelting industry should have dealt years ago with the problem of sulfur dioxide emission, this lapse by no means justified the near-hysteria that built up in 1970 and 1971 over environmental pollution. This crusade found one outlet, in Arizona especially, in a drive to curb smelter emissions that went far beyond the limits indicated clearly by reason or logic. The sequence of events, in relation to Magma, was as follows:

In May, 1970, the Arizona legislature enacted a law governing air quality in the state and also provided standards and regulations for control of various sources of pollution, including copper smelters. In addition to setting ambient air standards for sulfur dioxide, the regulations required removal of 90 per cent or more of the sulfur dioxide that would otherwise be emitted in the smelter's stack gases. The U.S. Environmental Protection Agency (E.P.A.) later set Federal standards that were much less stringent than these state standards and that did not include the 90 per cent standard.

On investigation, the Magma staff concluded that the most readily available means of meeting the Arizona standards was to replace the reverberatory furnaces by flash smelting furnaces, as developed by Outokumpu, Oy., the Finnish copper smelting company, and to treat the flash furnace gases and the converter gases in a sulfuric acid plant. Reverberatory furnace gases are too low in sulfur dioxide content to be fed to an acid plant, but both flash furnace and converter gases are rich enough to make suitable acid plant feed.

Preliminary estimates, completed in early 1971, indicated that the cost of flash furnace and acid plant facilities to control air pollution at San Manuel would be about $50 million. However, detailed engineering work, undertaken thereafter, began to show, by mid-1971, that the cost would be at least $88 million and might even run much higher. Furthermore, with the flash furnace-acid plant system, the smelter would produce, at full capacity, about 3,000 tons of sulfuric acid daily, for which no market existed. Disposal of this acid would be, in itself, a formidable technological and economic problem involving no likelihood of profit.

This was just too much money to spend on a project that involved little hope of profit, particularly when it became clear in August, 1971, that a 90 per cent elimination of SO_2 from the smelter gases was not needed in order to meet the Federal air-quality standards promulgated that month by the E.P.A. Therefore, Magma began studying other means of pollution control, but at the same time continued the flash furnace engineering work as a stand-by.

Magma, and other Arizona copper mining and smelting companies, tried diligently in late 1971 to convince Arizona's state agencies that the Federal air-quality standards provided more than ample protection for health and welfare. Some measure of success was achieved when in May, 1972, the state submitted to the E.P.A. a revised control plan that dropped the 90 per cent emission control standard and established ambient air standards approximating the stringent Federal secondary ambient air standards.

The E.P.A. rejected Arizona's plan, and in July, 1972, issued proposed regulations of its own for Arizona. These included a specific percentage reduction in SO_2 emission for each smelter. For example, the Magma smelter at San Manuel would be required to reduce its SO_2 emission about 70 per cent by mid-1975 (which Magma had by then planned to do by 1974 anyway). However, by a use of a formula based, as it turned out, on wholly inaccurate data, the E.P.A. proposed that San Manuel be required by

mid-1977 to capture 94.6 per cent of its sulfur emission in order
to meet only the Federal primary ambient air standards. A
further reduction to meet the more stringent secondary standards
would be called for in 18 months when the E.P.A. made up its
mind what the reduction should be.

Basing this outrageous proposal on patently erroneous data
was too much even for the State of Arizona, and as this was being
written, discussions were raging between the state and the E.P.A.,
hearings were being held, and lawsuits were being threatened by
nearly everyone concerned. A measure of the gravity of the
problem was that reliable estimates put the cost to the copper
industry of full compliance with the E.P.A.'s proposals at well
over $600 million, a staggering investment that would have to
be made with little or no hope of economic return.

Rather than submit tamely to what seemed such unreason-
able demands, the copper smelting industry braced for a battle.
On the one hand, the companies laid plans, and began construction, for measures designed to achieve a degree of pollution
control in line with Federal primary standards. On the other,
most companies, Magma included, prepared to take whatever
action was necessary to fight successfully the extreme requirements they regarded as ruinous and unwarranted.

Fortunately, as the enormity of the burden to industry became clearer, it began to dawn on many legislators that the resulting cost to the public and the national economy of such
extreme control measures might far outweigh the benefits. There
were, in mid-1972, some indications of returning sanity in the
environmental drive.

Malozemoff found this reassuring. He had been greatly
troubled by the fact that Magma's expressed desire to do all that
was necessary, but no more than was logically necessary, to clean
up the Arizona atmosphere had been brushed aside by the
crusaders, and that Newmont and Magma had been threatened
with ruinously uneconomic expenditures to be forced on them
for illogical, unjustifiable reasons. He felt as if his control over
the Newmont enterprise were under attack in a nightmarish assault that would not yield to reason. However, he believed also

This group met at Idarado Mining Company, Ouray, Colorado, in 1943 to check progress in driving the Treasury Tunnel. From left are: Roy Bonebrake, Newmont; Oscar Johson, Mine & Smelter Supply; a Mr. Nicholson; Johnny Edgar, Idarado manager; and Long John Austin, tunnel driver.

Resurrection mine near Leadville, Colorado, owned jointly by Resurrection Mining Company and American Smelting and Refining Company, and managed by the latter, began mining and milling 700 tons lead-zinc-silver ore daily in April, 1971.

Austin McVeigh, discoverer
Sherritt Gordon's Lynn La
orebody, points out the very o
crop from which he cut samp
in 1941 that assayed high in ni
el and copper and led to a wh
new enterprise.

Eldon L. Brown, for years Sherritt's chief executive, chips sample from high grade nickel-copper face in El orebody of Lynn Lake mine. This photograph was taken in November, 1953; Sherritt's chief emphasis has since been transferred to new mines in Manitoba.

Carlin Gold Mining Company's property in Nevada was designed for minimal ecological damage long before that problem became of public concern. All process water is recycled from behind this dam, none overflows.

Similkameen Mining Company in British Columbia uses three huge pebble mills for grinding its copper ore. Mills are shown awaiting installation of 38-ft. diameter drive gears. Mill is now in operation at better than planned capacity of 15,000 tons of ore daily.

Palabora Mining Company's open pit in South Africa produces about 60,000 tons of copper ore daily, but company turns out uranium oxide, magnetite, vermiculite and sulfuric acid as well as blister, wirebar, and continuous-cast rod copper. It is one of the most efficient producers in the world.

Atlantic Cement, on the Hudson River near Albany, N.Y., is enjoying (1972) profits after years of losses. Loading dock is in foreground, plant in rear, and conveyor from quarry in the background. *Glen S. Cook photo*

that there is still enough sense of proportion, or common sense, left in American institutions to make it impossible that an enterprise like Magma should be destroyed wantonly in a misguided effort to achieve an unrealistically pure environment.

In any event, Magma, believing that it must take some measure of pollution control action even though neither state nor Federal standards had been firmly established as governing, decided early in 1972 on an air pollution control program that was believed capable of meeting the Federal primary ambient air standards. It should be in effect by the end of 1973. The plan was approved by the Arizona Hearing Board, and work on it began at once. In the first of the plan's two stages, a sulfuric acid plant will be built at San Manuel to remove SO_2 from the converter gases, which will amount to a sulfur elimination of about 70 per cent. As a further control measure, eight monitoring stations have been established in the San Manuel area that will measure the SO_2 content of the ambient air and transmit their readings to the smelter. A computer will evaluate these data and signal a rising trend in time to allow curtailment in smelter operation, to avoid excessive emission of SO_2. Weather monitoring and forecasting stations are also provided.

This is the "closed loop" system of emission control that was developed and used successfully by the cities of Rotterdam, in the Netherlands, and Pittsburgh, Pennsylvania, and by Asarco at El Paso. Magma's system incorporates every improvement available, based on experience with these three systems.

In the second stage of the Magma plan, to the extent required, a wet limestone scrubbing process will be used to remove SO_2 from the reverberatory furnace gases. This process was still under test in mid-1972 at a pilot plant at McGill, Nevada, being operated by the Smelter Control Research Association, a nonprofit association formed by Newmont and seven other U.S. copper smelting companies to do research and pilot plant work on the removal of SO_2 and other contaminants from smelter gases. At the time of writing, reliable data had not yet been developed to enable design and construction of such a plant.

Contracts for the engineering, design, and equipment pro-

curement for the gas collection system and the acid plant were
signed by mid-1972, and the work was proceeding with all possible speed. Magma realized, however, that the question of pollution control was far from being resolved. The problem reaches
all the way from the atmosphere over San Manuel to the balance
of payments of the United States, and it may be years before this
chapter in the Newmont-Magma history can be completed.

The foregoing account of the merger with Magma and of the
Magma expansion program necessarily gives the appearance of
a smooth continuity that did not exist in actual fact. As Franz
Schneider had foreseen, the U.S. Department of Justice became
interested in Newmont's 80 per cent ownership of Magma, and
in 1962 brought suit against Newmont and Magma to abolish any
interlocking directorships among Newmont, Phelps Dodge,
and Magma and also to require Newmont to get rid of both its
2.9 per cent interest in Phelps Dodge and its Magma stocks.

In the late 1920's, Newmont had purchased a good deal of
stock in Calumet and Arizona and in New Cornelia Copper
Company, both Arizona copper mining operations. In the early
1930's, these companies were merged into Phelps Dodge, a move
in which Fred Searls and Henry DeWitt Smith played a large
part. The result was that Newmont acquired a holding in
Phelps Dodge, which over the years was built up to a 2.9 per
cent interest.

At the time of the suit, Franz Schneider, formerly a Newmont vice president, was a member of both the Newmont and
the Phelps Dodge boards. Also, Kenneth L. Isaacs, of Massachusetts Investors Trust, was a member of both the Newmont and
the Phelps Dodge boards. Plato Malozemoff and Roy Bonebrake
were also cited in the suit because of their membership on both
the Newmont and the Magma boards. Schneider and Isaacs vol-
untarily resigned from the Newmont board in April, 1963. Newmont could see its way clear to divesting itself of its Phelps
Dodge stock, but to give up the Magma stock was quite another
matter.

Following nearly four years of investigation and negotiation, the suit filed by the Department of Justice was settled by agreement in February, 1966. Newmont was left in possession of its stockholding in Magma, and no further change in directorship of either company was required. However, Newmont did have to divest itself of the Phelps Dodge stock within three years and could not thereafter have any director or officer in common with Phelps Dodge. Termination of the suit did not involve an admission of any kind on the part of any of the parties. The Phelps Dodge divestiture was accomplished by a dozen or so privately handled transactions, all of them through exchange of Newmont stock for Phelps Dodge stock with certain large shareholders of Newmont, and all based on prevailing market values. This operation had the approval of the Department of Justice and involved no tax either way.

Unquestionably, the Magma acquisition and resulting expansion was the most significant event in the Newmont history of the 1960's, but the company was active in many other areas as well. Fred Searls had always had a soft spot in his heart for the Resurrection mine at Leadville, Colorado, and he would be greatly delighted today if he could know that at long last, Resurrection is beginning to look like a solidly profitable operation.

However, back in 1953, things were not going too well for Resurrection. After the war and the end of the Premium Price Plan, Resurrection had struggled along, wavering back and forth between a small profit or a small loss, but falling metal prices in 1952 resulted in a loss that became even greater in the first half of 1953. At this point, Newmont, Hecla, and U.S. Smelting, the three owners of Resurrection, found themselves in deep disagreement as to what should be done with the property. Fred Searls had great faith in Resurrection, and when he realized that discussion had reached an impasse, he adopted the tactic he had used years previously in Idarado and offered to buy the interest held by Hecla and U.S. Smelting for $500,000 to each, or to sell Newmont's interest to them for a payment of $500,000 from each. Hecla and U.S. Smelting each promptly picked up $500,-

000 and retired from the scene, which dismayed Searls's associates at Newmont, who learned of Fred's offer only after it had
been accepted. Searls promptly stopped all mining at Resurrection, but continued exploration and development, although
some mining by leasers continued for years thereafter.

In 1955, Malozemoff negotiated an agreement between Resurrection and American Smelting and Refining Company by
which the Yak Tunnel was to be connected to the Irene mine of
Asarco, allowing the Irene property to be unwatered through
the tunnel. Under the agreement, the Irene mine and the Resurrection property in Iowa Gulch were to be operated jointly on a
50-50 basis, with the production being treated in the Resurrection mill. Although in the first half of 1957 about 90,000 tons of
lead-zinc ore of good grade was sent to the concentrator, a drop
in metal prices about mid-year put Resurrection back into the
red and the property was shut down in October.

After several years of inactivity Asarco and Newmont resumed exploration at Resurrection in September, 1961, but this
time the management of Resurrection was turned over to
Asarco, and it was operated thereafter by the western office of
that company under the direction of Frank Frost.

Frost rapidly developed an interest in the geology of the
area and spent a good deal of time and money for exploration,
particularly in zones much deeper than had been previously explored. Fred Searls, Bob Fulton, and Hal Norman of Newmont
contributed ideas to much of this exploration work. Little by little, the group developed about 2.4 million tons of ore averaging
5.13 per cent lead, 9.95 per cent zinc, 2.64 ounces silver, and
0.084 ounce gold per ton, which is high-grade ore by present-day
standards. Furthermore, it was felt geologically probable that additional ore reserves might be found deep under Iowa Gulch.

After considerable discussion between Newmont and
Asarco, it was decided to finance what has now become known as
the Resurrection Mine. The project involved sinking a new
shaft and building a new mill at a cost of approximately $14.7
million. Shaft sinking, which began in May, 1969, was completed at a depth of 1,654 feet by February, 1970. The job was

done by Harrison Western, Inc., on schedule and without a single lost-time accident. Designed by Stearns-Roger, the 700-ton-per-day lead-zinc flotation mill began operating early in 1971. Thanks to faith and perseverance, Resurrection seems to have developed as many lives as a cat, for it has now been resurrected at least a half dozen times since mining began there in 1879. Searls's confidence in Resurrection may yet be completely vindicated.

In southwestern Colorado, the Idarado Mining Company found itself in difficulties, as did Resurrection, at the end of the Premium Price Plan in July, 1947. Operations thereafter became unprofitable, and the management believed that the only feasible way of returning to profitability was to increase the tonnage treated. By 1949, Idarado had doubled production, but the extra earnings were largely taken up in buying additional mining claims and purchasing additional equipment. In 1950, Fred Searls, believing that Idarado needed still more funds, sold a 20 per cent interest to Homestake Mining Company, the gold producer of Lead, South Dakota.

John Wise aided Idarado greatly in the period 1952 to 1962. In his ten years as manager, production increased from 261,000 tons of ore per year to over 400,000, with a peak of 458,000 in 1957. At the same time, productivity rose from 5.25 tons per mine man-shift in 1952 to 8.16 in 1962. In 1952, total operating cost was $11.69 per ton mined; in 1962, it was $8.45. Wise and Bob Hilander, mine superintendent, wrought this near miracle by unrelenting supervision of hiring, training, distribution, and performance of men and close control of equipment and supply costs. Idarado was one of the first U.S. mines to use throwaway rock bits. Fred Wise introduced them in the late 1940's, and John Wise improved their performance.

In 1962, Plato Malozemoff wrote Banghart a note about Idarado saying, in part: "The cost and productivity record through the years is really remarkable, and Johnny Wise and Bob Hilander deserve all the credit for this work."

Arthur C. (Bob) Hilander was born in Chicago, but went

to the University of Alaska for his higher education, receiving his B.S. degree in mining engineering there in 1942. He served with the U.S. Air Force during the war, and was discharged in 1945 with the rank of captain.

Thereafter, Hilander joined Telluride Mines, Inc., and became an employee of Idarado when it purchased Telluride Mines in 1953. When John Wise moved to New York in 1965, Hilander became general manager of Idarado. At the time of his death in April, 1971, as the result of a brain tumor, he was general manager of Carlin Gold Mining Company and Dawn Mining Company, as well as Idarado. Widely known for his friendliness and keen interest in community and state affairs, Bob Hilander left an enviable record as a mining engineer and a good manager.

Idarado's future really brightened in 1953 when, as a result of an idea of Fred Searls's, Idarado bought all the common stock of Telluride Mines, Inc., for $1 million cash, in a deal negotiated by Fred Searls and John Wise, to consolidate the claim ownership of the two companies, which were located on either side of the Uncompahgre Range. To get from the Red Mountain office of Idarado around to the Pandora Mill office at Telluride, one has to drive about 70 miles, although he could get from the one to the other directly through the mountain by traversing about 5 miles of mine workings, not all of them on the same level.

Buying Telluride Mines, Inc., thus put Idarado in possession of famous old mines like the Tomboy, Smuggler, Pandora, Liberty Bell, Argentine, and others. Idarado now owned two mills, neither of which could be regarded as a modern and efficient producer. Studying the Idarado problem was one of the first assignments handed M. D. Banghart when he returned from Africa to New York in 1954 to replace Henry Smith, who was retiring as vice president of operations of Newmont. Almost at once, Banghart reached the conclusion that the Red Mountain mill was in the wrong place from the combined Telluride-Idarado point of view. Considered horizontally, vertically, and geographically, the

Telluride location was a better mill site than that of Red Mountain. Following considerable discussion, largely held among John Wise, Frank McQuiston, Banghart, and Plato Malozemoff, Newmont decided to shut down the Red Mountain mill and to modernize and expand the Pandora mill at Telluride. Banghart and Wise then initiated a program of mill modernization and mine development that, in all, took about five years and cost around $2 million. The hardest part was an eighteen-month program of connecting various levels to improve ventilation and to enable all ore to move down and out at the Pandora mill tunnel level. All this was financed out of earnings plus a bank loan to Idarado.

By 1959, when the program had been completed and the new Pandora mill at Telluride was in production, Idarado had become a much more efficient and a much lower-cost operation. For years Idarado milled about 400,000 tons of ore annually, running about 0.05 ounce of gold, 2.25 ounces of silver, 2.5 per cent lead, 3.75 per cent zinc, and 0.75 per cent copper. Through 1970, Idarado paid $18,262,568 in dividends, but in 1971 a combination of low metal prices, rising cost, a severe labor shortage, and a strike at the smelters reduced earnings to a point where no dividend could be paid in that year. This situation had begun to improve in 1972, however, and Idarado has resumed paying dividends.

As mining proceeded downward on the vein systems in the Idarado claims, the miners encountered a new type of ore occurrence that has greatly enhanced the property's promise. In the old workings, near the top of the mountain range, the veins were rich in gold and silver, but a zoning developed at depth in which the zinc increased and lead, gold, and silver decreased. Beginning in the San Juan tuff, at the top of the mountain, veins come down through the Telluride conglomerate below the 10,000-foot level, and before reaching the Pandora mill level at 9,000 feet, the veins enter the Red Beds characteristic of the Ouray and the Telluride area.

All the way down through the San Juan tuff, the veins, such

as the Black Bear and the Argentine, though steeply dipping, exhibited the characteristics of a normal vein system. However, on entering the Telluride conglomerate below 10,000 feet, the veins, in some places along the strike, suddenly widen into zones containing replacement-type orebodies, enabling mining to be done in stopes 30 to 40 feet wide. This discovery, though not quite as dramatic from the point of view of increased ore reserves as the discovery of the replacement orebodies at Magma's mine in Superior, still provided Idarado with added higher-grade ore reserves and an opportunity to mine at somewhat lower cost. Search for more of this replacement ore continues.

"Fortune favors the brave" could have been said of San Manuel, Idarado, and in its most recent development, even Resurrection. However, it is necessary to record one Newmont venture of the 1960's that was begun bravely enough, but that has been ill-favored, until recently, not only by fortune, but in several other aspects as well. This was Newmont's venture into the cement industry, which began in 1958 when a Swiss engineer, experienced in the cement industry, approached the Cerro Corporation with a proposal to establish a cement plant on the eastern seaboard.

It was not to be an ordinary cement plant, however. The engineer had observed that traditionally the size of a cement plant was limited by the size of its surrounding market area, generally considered to be within a 150-mile radius as the maximum economic range for land transportation. He thought, quite correctly, that operating- and marketing-cost economies could be achieved by building a really large cement plant in a location where access to low-cost water-borne shipment would greatly expand the plant's potential marketing area.

Such a location existed along the Hudson River, where large limestone deposits were known to exist close to the river, and from which cement could be moved by water all up and down the Atlantic Coast. Upon investigation, the proposal looked attractive for several reasons. Offered to Newmont by

Cerro as an enterprise in which each company would share equally, the venture in cement represented a diversification for Newmont and, furthermore, would mean an addition to Newmont's proportion of domestic investment. Also, it appeared to offer a means of reducing production and distribution costs in supplying the building industry at a time when the outlook for construction throughout the country appeared excellent.

Cerro and Newmont then authorized intensive cost analyses and market surveys to make as certain as possible that the project was feasible and potentially profitable. Two consultants were employed, who went about among cement customers, practically door-to-door, getting their opinions on cement requirements and prices. Their report on prices gave the opinion that prevailing prices would hold. Actually, prices dropped after the start of operations, and it was eight years before prices again reached the levels prevailing at the time of the study.

In making the economic analyses, the consultants carefully worked out the economic effect on the project of a drop in the cement price, or of a shortfall in production of cement, or of a sharp rise in production costs. Apparently, everyone ignored what the effect would be if all three unhappy events took place all at once, which was unfortunate, because that is just what happened.

However, on the basis of the studies, Cerro and Newmont announced in 1960 that each company had taken a 45 per cent interest in the newly formed Atlantic Cement Company, Inc. The Swiss engineer and several of his partners took the remaining 10 per cent. From a quarry and plant located on the Hudson River at Ravena, New York, Atlantic Cement was to produce up to 10 million barrels of cement annually. Cost of the project, which was to be completed in 1963, was an estimated $64 million. From Ravena, the cement was to be shipped by water in the largest self-discharging barges ever built, which carried loads of up to 17,000 tons each to ten or more distribution points, from which the entire east-coast market for cement could be served. The most distant point was 1,500 miles away. At 10 million bar-

rels per year, Atlantic Cement would have one of the largest cement plants in the world. Limestone reserves near Ravena were
estimated to be adequate for about one hundred years of operation at that level.

Atlantic Cement financed the project by a $33 million loan
from an insurance company and another loan of $14 million
from a group of banks. The rest of the $64 million investment
came from Newmont and Cerro in equal shares. Largely to
avoid any possible conflict over management, Newmont and
Cerro, by agreement, left direction of the project at first in the
hands of the Swiss engineer and his group. After several disappointing years, the group was bought out in 1967, leaving Newmont and Cerro with 50 per cent ownership each. Thereafter,
by mutual consent, Newmont took the lead in management.

Although construction costs did not exceed the original estimate, it was discovered later that this seeming success had been
obtained at considerable sacrifice in equipment capacity and
general plant design. The result was that production fell woe-
fully short of the planned 10 million barrels. The highest annual output occurred in 1971, when the plant produced 8.2 million barrels for the year. Actually, Atlantic has a solid
8-million-barrel plant. It is unlikely ever to become a 10-million-
barrel plant without major, and expensive, additions.

The major reason for this continuing shortfall lies in the ro-
tary kilns, 580 feet long and 20 feet in diameter, the largest cement kilns ever built when the plant was designed in the early
1960's. They were optimistically rated at 5 million barrels each.

This overrating is not surprising when one reflects that no
one really had any accurate idea of how a kiln that size would
act or what maintenance would be required. An ordinary cement kiln is available for production 85 to 95 per cent of the
time. Atlantic Cement's kilns were operable only 70 per cent of
the time, on the average, largely because of down time for maintenance on the interior brickwork. Frequent shutdowns caused a
thermal shock that had a disastrous effect on the big kilns. The
key angle in the large diameter shells is so flat that repeated expansion and contraction loosened the brick.

Also, the slurry goes through the kilns very slowly, taking about four hours to pass through. During the trip, the material goes through a number of physical as well as chemical changes, and the effect of a control error may not be felt for a couple of hours after it occurs. All this was not well understood at first, but not even full understanding could add a million barrels to each kiln's capacity. However, Atlantic Cement's kilns in 1971 operated 82.6 per cent of the available time, a strong improvement over the poor record of the early years.

The effect of this shortfall in kiln capacity, plus other inadequacies of design, came on top of a slackening in demand for cement and a weakening in price that lasted until 1970. Consequently, Atlantic Cement recorded steady and severe losses through the 1960's. Along in 1966, Newmont and Cerro began a program of rehabilitation aimed at increasing production and cutting costs. As noted above, they also bought out the other partners. The major elements in this program were:

1. Adding a fourth grinding mill, which had been dropped from the original design.
2. Conversion of grinding mill twin drives to a conventional type of drive with a single motor.
3. Establishment of local captive supply sources for all raw materials, and provision for improved handling and storage of the raw materials.
4. Conversion from coal to fuel oil as the energy source.
5. Installation of a bag house to control clinker cooler dust.

All these were seven-figure investments, but all except the last one had a significant and beneficial effect on production and costs. Also, for years, Walter Schmid, who was the Newmont treasurer in 1969, had been advocating a transfer of Atlantic's debt to the parent companies. In late 1969 he was successful in convincing both Cerro and Newmont that this should be done. In December, 1969, and January, 1970, Newmont and Cerro each acquired from Atlantic's lenders equal shares of the mortgage bond and notes previously issued to them by Atlantic. As a result, Atlantic has been relieved of its indebtedness and the consequent payment of about $3 million in interest annually. Newmont and Cerro each assumed an indebtedness of $21,-653,000.

Atlantic reported net income in 1970 of $1,105,000, but in 1971, net income rose to $7,541,000, which included nonrecurring gains of $2,022,000. For the first time, Atlantic was able to pay a dividend in 1971, amounting to $2 million to each parent company.

At present, Atlantic is directed by an enthusiastic management team headed by Donald M. Halsted, Jr., president, and Charles S. Burriss, vice president, operations. With the plant running normally at a higher rate than 8 million barrels annually and with demand growing and prices firm at levels better than those anticipated eight long years ago, these men look forward to a future that seems likely to justify the hopes that were held for Atlantic when it was conceived.

Newmont's drive in the mid-1960's to acquire additional sources of income in North America has resulted in still another copper producer in British Columbia, now entering production under the name of Similkameen Mining Company Limited. Like many another Newmont venture, Similkameen began with a casual conversation in late 1965 that led to an examination by Newmont of a prospect held by Jerry Burr, a garage owner in Princeton, British Columbia. The prospect was in the hill just across the Similkameen River that runs through Princeton. It was just opposite a copper property called Copper Mountain that was owned, and had been mined by, Granby Consolidated, but had been shut down since 1957.

When Don Cannon, a Newmont geologist, visited the prospect, he thought pretty well of it and called John Drybrough to say that it looked good. After further examination, a deal was put together in January, 1966, that brought Burr's prospect into Newmont's hands. Thereafter, this prospect, and the orebody that was developed from it, has been known as Ingerbelle.

Because it seemed appropriate to put the Ingerbelle and the Copper Mountain orebodies together, Newmont approached Granby regarding the latter property. Newmont eventually bought Copper Mountain in 1967 for $11.7 million in cash and stock equivalent.

Having acquired both properties, Newmont set about delineating ore reserves. Drilling was done under the direction of John Livermore, who worked so effectively with Bob Fulton on the exploration in Nevada that turned up the Carlin gold mine. (The Carlin story is told in the next chapter.) Drilling at the Similkameen-Ingerbelle property, was unusually difficult because the formations are rather badly broken up by complex faulting, and mineralization is often erratic. Of the first sixteen holes drilled, six showed good copper content, but the other ten were quite poor. Also, at one point, in looking over the drill map, Drybrough observed an almost completed ring of holes which still contained a gap of about 400 feet without a hole. Drybrough asked Livermore to drill in this gap because he thought "what would New York say if we didn't have a hole there?" The result was the best hole on the property.

The Ingerbelle orebody is on the west side of the Similkameen valley, which runs northeasterly, and the old Copper Mountain orebody is on the east side. Total ore reserves are estimated at 76 million tons at 0.53 per cent copper. Steel erection for a 15,000-ton-per-day-capacity mill began on December 1, 1970, under the direction of Canadian Bechtel Limited. Design of the mill was supervised for Newmont by Frank McQuiston and Dave Christie, and was carried out by the Bechtel Corporation in San Francisco. The mill will work on ore from the Ingerbelle orebody during the first eight years of the life of the property, and beginning about 1980, it will have another seven years' life on ore from Copper Mountain. Initial production began in March, 1972, and full production was attained by September.

The man in charge for Similkameen at Princeton from the time Newmont bought the Copper Mountain orebody is Harvey Parliament, formerly an employee of Granby Consolidated. As executive vice president of Similkameen Mining Company

Limited, Harvey has guided the project from the drawing board stage up through full production. The fact that Similkameen entered production ahead of schedule, within the capital expense budget, and at 17,000 tons of ore daily instead of the rated 15,000 tons owes much to Harvey Parliament's leadership.

Total cost of the venture, including the drilling campaign and the purchase of the Copper Mountain orebody, is estimated at $73 million. Of this amount, $26 million was advanced by Newmont, a loan of up to $9 million was obtained from the Export-Import Bank of Washington, and $38.7 million (Canadian) was obtained as a loan from the Canadian Imperial Bank of Commerce. Details of the financing, as arranged by J. O. Cumberland, appear in Chapter 16.

Progress in mining in the past fifteen or twenty years is pointed up by the fact that the Similkameen operation is expected to be reasonably profitable even though it will be working on ore that was regarded as quite marginal back in 1957, when Granby stopped mining at Copper Mountain. A major difference between 1957 and 1972 lies in the fact that in 1957 Granby was using 1.5-cubic yard shovels and 10-ton trucks, whereas in 1972, Similkameen will be using 10-cubic yard shovels and 100-ton trucks. The increased equipment size and higher production rates are major contributors to lower-cost mining.

Copper concentrates from the mill go to Japan under a long-term purchase contract. Mill tailings, the disposal of which is so expensive a problem to so many mining operations, was a special problem at Similkameen. Tailings go across the valley on a suspension bridge to be deposited in a dead lake, for which there is no outlet, and which was dammed at both ends to prevent discharge as tailings are deposited. The lake is occupied only by algae and leeches. Similkameen will thus avoid water pollution in the Princeton area.

In this record of financing new ventures at home, the five projects described in this chapter have involved Newmont directly in the financing of a total investment of over $580 million. In addition, Newmont has another $83 million or so of its funds invested in enterprises that are only in the development or early production stage, such as Vekol Hills in Arizona, or in such companies as Highveld Vanadium in South Africa or Foote Mineral Company, which have not yet made significant contribution to Newmont's earnings. Thus, over $660 million in economic enterprise is traceable to the change in Newmont's philosophy that began with the $13 million investment in Sherritt Gordon back in 1952.

There can be no quarrel with the statement that Newmont possesses a cool and competent financial judgment. Yet financial judgment is only one leg of the tripod in the mining industry. Technological competence and skill in exploration are the other two. Because Newmont has been quite successful at exploration, it seems appropriate to devote the next chapter to a discussion of the modern methods of exploration that it has developed.

Exploration Party, Alaska

14 Newmont and Exploration

Without successful exploration or acquisition, a mining company can live only as long as its immediate ore reserves will last. Obviously, to have grown as it has in the past fifty years, Newmont Mining Corporation must have done a markedly successful job of exploration in the broadest sense, which involves an open-minded, imaginative approach that goes beyond discovery to successful extension or enlargement of ore reserves by intelligent acquisition. This chapter examines how Newmont's geologists and geophysicists went about doing this job and illustrates their technique with the example of the successful exploration program that resulted in the development of Carlin Gold Mining Company in Nevada.

Mention of the word "exploration" is likely to create in one's mind a picture of an intrepid prospector pushing out into a desert or an unexplored mountain range stubbornly searching for an outcrop or some other surface indication of an orebody that may make him or his company rich. Such prospectors and such discoveries still exist. In fact, no better example could be found than Austin McVeigh, the prospector who came across the outcrop that has blossomed into the new Sherritt Gordon enterprise. However, new discoveries of this nature are distinguished more by their rarity than by their frequency. Since 1900, when W. B. Thompson became interested in the Shannon mine in Arizona, Thompson himself, and later Newmont, have been involved in ninety-three major projects or investments in the mining and petroleum fields. Very few of these have been the result of entirely new discoveries on the part of a Newmont geologist.

Actually, the vast majority of Newmont's producing properties have been developed successfully because Newmont's exploration men saw something that other people had missed, or because they were able to provide the owners of a prospect an avenue to New York for financing the project, or because an advance in technology had made certain deposits commercial that previously had been worthless. Successful exploration is really as much a matter of evaluation as it is a matter of discovery.

Would it be accurate to call Henry Krumb an outstanding "exploration" geologist? Probably not. Although Newmont owes many of its profitable acquisitions to Henry Krumb's ability, he was noted for a scientific approach to evaluation of mineral properties rather than to the kind of field work that led to discovery. On the other hand, Fred Searls, who established Newmont's first exploration staff, was undoubtedly one of the best field geologists in the industry. Searls, Krumb, A. J. McNab, and later, Henry DeWitt Smith made an excellent exploration team in that their special abilities complemented one another. Searls and his field staff would bring in a prospect and if it looked promising, Krumb and McNab or Henry Smith would "shake it down" to determine its commercial attractiveness and make realistic plans for its production phase.

When Fred Searls joined Newmont in 1925, he knew that if Newmont were to grow, or even to survive, a vigorous exploration program had to be developed. In addition to bringing in some of his old friends, as described in earlier chapters, Searls himself was quite literally all over the world in those days. His tireless search for prospects covered practically all of Africa, especially central and southern Africa. Old Roman gold mining properties in Spain came under Fred's observation, as did prospects in Russia, China, Turkey, and Iran, as well as a great many prospects in Canada and the United States.

In the course of this activity, Fred Searls developed a definite philosophy of exploration for Newmont, and the basic principles remain unchanged to this day. For example, when it came to considering a prospect, Searls would never accept second-hand information. He insisted on having someone from Newmont, if not himself, do the necessary geologic mapping and field investigation, even though such work might duplicate that which had already been done by someone else. Searls felt strongly about this, and inculcated this attitude in each new geologist who came with the company.

Although Newmont's widespread exploration activities required the services of many geologists, Searls endeavored to keep their number as small as possible. Newmont has never operated a world-wide network of exploration offices, as some other mining companies have done. For news of prospects, Searls depended on his numerous friends in most of the world's centers of mining activity and on numerous listening posts in various other parts of the world.

It was Searls's belief that a large and expensive staff may very well devote itself to finding things to justify its existence, rather than to concentrate on weeding out unacceptable prospects. In other words, Searls believed that a small, hard-working staff is less likely to be under pressure to produce, and is therefore less likely to take long chances and possibly make expensive mistakes.

As examples of the importance to Searls and to Newmont of friendships, one may cite Idarado and Dawn. Searls became acquainted in 1936 with an opportunity at Idarado through his friendship with Oscar Johnson, president of the Mine and Smelter Supply Company of Denver, and with Spencer Hinsdale, a banker in Portland, Oregon. Nearly twenty years later, it was Spencer Hinsdale who called Fred Searls and told him of an opportunity for Newmont that was arising in the uranium property in the State of Washington that was eventually developed by Dawn Mining Company.

Having seen this philosophy of exploration, at first hand, working both successfully and inexpensively, Plato Malozemoff and Robert B. Fulton, who was Newmont's vice president for exploration until June, 1972, have maintained this exploration policy, making only such changes as have been necessary to keep the program in line with modern exploration techniques.

Bob Fulton, now president of McIntyre Porcupine Mines, Limited, of Toronto, was born in the heart of the Mother Lode country in California in the small town of Melones in Calaveras County. Graduating in mining engineering from the University of Nevada in 1941, Fulton spent not quite a year with Consolidated Copper Mines in Nevada before he entered the Navy. After three years in the Navy, Fulton returned to Nevada and found a position with the Natomas Company, which was setting up a gold dredge near Battle Mountain. At the end of 1946, Fulton went to work for Calaveras Cement Company, one of the William Wallace Mein enterprises, but in July of 1947 he joined the Empire Star Company as a mining engineer. From the position of superintendent of the North Star mine, Fulton was moved to the New York office of Newmont in October, 1950 as office engineer. His first assignment in New York was to work with Fred Searls in going over a great number of property reports, most of them on porphyry copper prospects, with a view to unearthing opportunities for using the new geophysical technique that Dr. Arthur Brant was developing.

In 1956, Fulton was sent out to Washington as manager of the uranium operations of Dawn Mining Company. He remained there until early 1958 when he was brought back to New York as manager of exploration. Fulton became vice presi-

dent of Newmont in 1961 in charge of the company's exploration activities in the U.S. and Canada.

Of vital importance in Newmont's exploration program is
the science of geophysics. Fred Searls was no stranger to this new
technique. His interest and enthusiasm initiated an inquiry that
led to a major Newmont innovation in geophysics, the so-called
Induced Polarization method, or IP, for short. In his work with
the War Production Board during World War II, Fred heard of
some new techniques having to do with spotting submarines
that he thought might be useful in spotting orebodies. He
talked these ideas over with Henry Smith, Phil Kraft, and Franz
Schneider, who agreed that Newmont should try to adapt these
techniques to mineral exploration.

After about a year of unsuccessful experimentation, Searls
engaged Dr. Arthur A. Brant, and the research began to show results. Born in Toronto, Dr. Brant studied mathematics and
physics at the University of Toronto, Princeton University, and
the University of Berlin, where he obtained his doctorate in
geophysics in 1936. An out-of-doors man, Dr. Brant played
championship hockey during college and coached a German
national hockey team while studying in Berlin.

Dr. Brant taught physics as a professor at the University of
Toronto until 1948, during which time he served as geophysical
consultant to several Canadian mining companies. In 1948, he
was retained by Newmont, and in 1949 he became a Newmont
employee and is now director of the Newmont geophysical department, headquartered in Danbury, Connecticut.

A prolific author and lecturer, Dr. Brant also holds, or was
co-inventor in, fourteen patents in geophysical instrumentation
and applications. He has won a number of gold medals and special honors, such as The American Institute of Mining Engineers' Jackling Award, yet his approach to people is unassuming
and unselfconscious. On meeting Dr. Brant, your first impression is of a big hand and a strong grasp. Your second impression
is of a man who appears to be more interested in your opinion

than in his own. Dr. Brant welcomes new ideas and new knowledge wherever and however it may come to him.

Along with retaining Dr. Brant, Searls commissioned the Radio Frequency Laboratory of Boonton, New Jersey to investigate geophysical use of wartime techniques. The Boonton lab drew attention to the phenomenon whereby metallic particles in a water tank, after passage of a direct current through the water, gave rise to an electrical discharge of considerable duration. In other words, the metal in the water acted as a capacitor, first storing up and then discharging an electrical charge. This was the basis of a metal mine detector used by Navy frogmen during the war.

Dr. Brant recommended that this characteristic be further investigated, because if metallic particles in water behaved this way, possibly metallic mineral particles in an orebody would do likewise. The Boonton lab ran field tests at Ely, Nevada, in 1947 and repeated them with Newmont at San Manuel in 1948. These tests, done over known porphyry copper orebodies, showed that disseminated sulfide mineral particles would respond as did the metallic particles in water. The resulting electrical discharges could be detected even under a great thickness of rock cover, particularly if the mineral particles occurred in large volume, as they do at San Manuel.

Thus began development of the Induced Polarization (IP) method, which is the first geophysical technique capable of detecting scattered or diseminated sulfide mineralization. Developed and used at first only by Newmont, the method has now been adopted by many other mining companies and independent geophysical contractors throughout the world.

Largely because of its location in the porphyry copper country of the southwestern United States, Searls established in 1948 a geophysical laboratory at Jerome, Arizona, with Dr. Brant in charge. This marked the beginning of a coherent and continuing geophysical effort by Newmont, still carried on under Dr. Brant's leadership.

Emphasizing the rapid growth of opportunity for young men in the mining industry, the Jerome laboratory employed several young men knowledgeable in physics and electronics. With the interest and skills they developed at Jerome, these men have gone on to become consultants in their own right, or to head geophysical services of their own.

A successful use of the new IP technique was at Cuajone, Peru, between 1952 and 1955, in the work preliminary to the establishment of Southern Peru Copper Corporation. At Cuajone, the geophysical work relatively inexpensively expanded the limits of previously known mineralization and permitted a more economic use of subsequent drilling.

Among other innovations, the Jerome laboratory developed the first electromagnetic airborne system, in which both the transmitting primary coil and the receiving coil were rigidly attached to the aircraft. This was first incorporated in a Sikorsky S-55 helicopter. Aeroservice Corporation of Philadelphia was the first to use this method in a job for Newmont, and later Aeroservice used it for others. This method permitted flying at no more than 150 feet above ground, and it yielded a sensitivity of response never before obtained. Among others who have used this new technique successfully are Amax and International Nickel, the Heath Steele copper-lead-zinc orebody in New Brunswick having been found in this way.

Testing the new IP method by Newmont in Africa was a natural development. However, at Tsumeb, the IP method simply did not work. Two limestones, one containing pyrolusite and the other carrying carbon, gave an IP response over practically the entire Tsumeb area. Similarly, the irregular magnetite-bearing Nababeep gneiss at O'okiep gave a misleading response.

However, geophysical methods were not totally unsuccessful at O'okiep and Tsumeb. Down-the-hole electrical resistivity and IP work was done in several O'okiep drill holes. The work at Nababeep West indicated the steep downward trend of the upper narrow sulfides at just about the same time as geologic interpretation arrived at an awareness of this condition.

Routine magnetic work began at O'okiep in 1950, and it has been supplemented by gravity methods since 1956. Because three of the host rocks of the region are magnetic, magnetics are used for reconnaissance and later localization of suspected orebodies. For example, the drilling at Spektakel was largely guided by magnetic and gravimetric work done around the known outcrop. Also, recent airborne magnetic work at O'okiep has suggested a greater frequency of outlying basic rock occurences than had hitherto been expected.

By 1957, Newmont's activity in geophysical exploration had grown so large that the company decided to handle field coverage thereafter by contract wherever services were available and adequate. Also, Newmont's new research laboratory at Danbury provided a more useful location for the geophysical development group. After moving to Danbury, the group has concentrated on interpretive techniques and the development of special methods in geophysics. From Danbury, the group has made material contributions in interpretation of observed phenomena at Ingerbelle, in British Columbia, and, in Arizona, at Vekol Hills, and at the Copper Creek prospect.

Instrument development is an important activity at Danbury. George McLaughlin has worked out an IP receiver that can be operated without interconnection to the generating or charging circuit. This permits several receivers to be operated simultaneously and thus provides faster and more economical ground coverage. McLaughlin has also developed a compact miniaturized unit for measuring physical properties of host rocks. The independent receiver is now marketed by various instrument fabricators, and the physical property instrument will also become commercially available in the near future.

It is Dr. Brant's conviction that as direct exploration targets diminish in number throughout the world, a greater reliance must be placed on evaluation of extensive geophysical data, and more effective means must be provided to test subsurface rocks economically. With the guidance of imaginative geology, Dr. Brant's group hopes to achieve this goal of more effective and

more economical testing through a continuing process of research and development.

For an expression of Newmont's general philosophy of exploration in action, no better example could be cited than the
discovery of the Carlin gold orebody in Nevada. In a word, Carlin is different. To locate it required a new approach to gold
mine prospecting. To mine and to mill the Carlin ore required
still more new approaches. To bring the property into production called upon every aspect of the multifaceted Newmont competence.

The Carlin venture began for Newmont when a geologist of
the U.S. Geological Survey named Ralph J. Roberts wrote a
short paper, only two-and-one-half-pages long, as a part of a
collection of papers published in 1960 by the U.S.G.S. as "Professional Paper 400-B." No doubt a great many geologists with
various mining companies saw this paper, but apparently only
Bob Fulton and John S. Livermore, a Newmont geologist, carried the study of it to a logical and useful conclusion. Briefly,
what Roberts and other geologists of the U.S.G.S. had done over
a thirty-year period was to map a large area in north central Nevada, an area much larger than any single mining company
could have covered. Regional exploration like this is not usually
attempted by mining companies because of the uncertain return
for the large expenditure of time and money required.

Roberts's paper made clear that in the region under consideration, there had been three or more large overthrust faults
that had resulted in great thicknesses of clastic sedimentary and
volcanic rocks being pushed eastward over younger marine formations. These overthrust blocks still exist throughout much of
the region as mountain ranges, but here and there erosion has
carried away a few "bubbles" of the older rocks to reveal the underlying carbonate rocks through what Roberts called "windows." These windows were often associated with igneous intru-
sives which could have provided the structural settings and the
mineral-bearing solutions that create orebodies. In fact, in some
of the windows there existed mineral deposits of commercial

value. Except for the windows, the underlying carbonates are buried by thousands of feet of the older overthrust rocks.

In considering Roberts's paper, Fulton and Livermore noted that his map depicted several windows in which were located earlier gold mining operations. They concluded that an exploration program based on some of these extensive windows might have a better chance of finding interesting gold accumulations than had earlier, unguided exploration efforts. Fulton took the paper to Malozemoff, who was intrigued by the overall concept and authorized an exploration campaign.

It may be noted that the objective of this search was, specifically, a gold orebody near the surface that could be mined cheaply, inasmuch as rising costs and the fixed price of gold had rendered deep underground gold mining unattractive for most companies.

For the exploration, John Livermore came over from Newmont's dormant venture at Ruby Hill, Nevada, and brought in a young English geologist, J. A. Coope, then working in Alaska. Together with Roberts and other U.S.G.S. geologists, the group went over the ground and laid out a prospecting program that took in the Marigold and the Buffalo Valley properties, near Battle Mountain, and the Goldacres, the Gold Quarry, and the Lynn Creek areas north of Carlin. The Carlin "window" was specifically targeted by Livermore.

The program in the Lynn Creek area was discussed in Carlin in October, 1961, at a meeting attended by the several geologists, plus Fulton and Fred Searls from New York. The group decided to stake a number of claims on open land within the Carlin window, following which a campaign of geologic mapping, trenching, and outcrop sampling was undertaken. Work was suspended during the winter and resumed in the spring of 1962, and before long, the company felt confident enough to acquire certain other properties in the area. One of these, which was held under a three-month option, was the first to be drilled. The third hole on this property cut 80 feet of rock assaying more than one ounce of gold per ton, comfortably close to the fabled "forty-dollar rock" of the Gold Rush days. The remarkable part of this discovery was that the gold-bearing zone began a little less than 10 feet below ground surface.

With more land acquired and more drilling done, the pattern of the deposit began to emerge. Drilling on a 100-foot-square grid pattern (Henry Krumb would have approved) continued until the end of 1963. By that time, about $350,000 had been spent, and Newmont knew it had a gold mine. Proven ore reserves eventually totaled 11 million tons assaying 0.32 ounce per ton, less than the indicated grade of that discovery hole, but still attractive.

The occurrence of gold in the Carlin orebody is so unusual that geologists are still debating the genesis of the deposit. Fulton believes that the gold is of magmatic origin and was introduced in solution probably as chloride. In a recently published study,[1] U.S.G.S. geologists Radtke and Scheiner suggest that the gold, entering the Roberts Mountains formation in solution as a chloride, encountered the carbonaceous materials in those rocks and was retained therein, possibly in part as colloidal gold particles adsorbed on activated carbon, but more probably as one or more gold organic compounds, which the authors suggest could have formed by reaction between the gold chloride and organic acids known to be present in these rocks. The exact nature of these gold organic compounds is still under study, but there is much experimental evidence to show that they exist. Little or no free gold is visible in these rocks, even microscopically, and the gold content cannot be adequately recovered by cyanidation.

Where this gold occurrence has been accessible to surface waters through exposure in the "windows", the carbonaceous or organic matter was destroyed by oxidation, and the gold was deposited in its native state, albeit in particles of sub-micron size, which for all practical purposes are invisible. This is the form in which the gold occurs in the Carlin Mine, and it can be recovered by cyanidation.

[1] "Studies of Hydrothermal Gold Deposition (I)," Arthur S. Radtke and Bernard J. Scheiner. *Economic Geology*, March-April, 1970, Vol. 65, No. 2, p. 87.

At the extremities of the Carlin orebody, where the organically-locked gold has been protected by overlying rocks and not subjected to oxidation, recovery of gold by cyanidation decreases accordingly. Carlin and Newmont metallurgists at the Danbury laboratory have developed a procedure, based on some work done by the U. S. Bureau of Mines, to recover gold associated with carbonaceous matter, and thereby enable mining and treatment of carbonaceous ores that could not otherwise be mined economically. In this method, carbonaceous ores are put through a pretreatment plant in which chlorine is used to oxidize the carbon, after which the gold can be extracted by cyanidation. Such a plant has operated successfully at Carlin on about 500 tons of carbonaceous ore daily since January, 1971.

In order to check the surface drill hole sampling results, bulk sampling of material taken from underground workings was carried out. These samples showed that the gold content of the orebody was not uniform. Furthermore, there were no visual mineralogical indicators that could be depended upon as ore guides in mining. Constant sampling for grade control was, and still remains, an absolute requirement.

A further complication arose in that during metallurgical test work on the Carlin ore, the Danbury staff discovered that the ground ore, from which cyanide had dissolved the gold, would not settle properly in the solution-solids separation process known as thickening. If thickening could be done only at prohibitive cost, utilization of the cyanide process was not practical. In a series of tests, the Danbury engineers found that by raising the alkalinity of the solution to a pH of 11, a chemical called Separan would induce proper thickening, whereas at the normal pH of 8, Separan did not work. A minor difference, but a critical one.

A third difficulty at Carlin was the special problem of preventing any escape whatsoever of cyanide-bearing solutions from the property. Even before the present outcry over environmental damage, the Newmont management had decided that the Carlin operation would be so designed as to create a minimum of disfigurement to the landscape and to contribute no pollution at all to the area's water sources.

To eliminate any plant discharge of water, Carlin built a large earth and rock dam in a canyon below the mill. All mill tailings are impounded behind the dam, and the only escape from the dam is by evaporation. All the retained water is recycled to the mill and there is no overflow. There is a special safeguard, however, whereby if overflow were to occur, cyanide would be automatically destroyed by injecting chlorine into the solution.

Construction of the Carlin mill and development of the mine began in the spring of 1964. On May 27, 1965, Newmont announced completion of the project at a total cost of about $10 million. At the dedication ceremony, Malozemoff said, "The Carlin gold mine vindicates our faith in this nation's ability to continue to find and to develop new mineral resources that are so vital for our well-being."

In Carlin's open-pit mine operation, broken ore is loaded on 35- and 65-ton trucks by means of 3- or 5-cubic yard diesel-electric shovels and hauled to the mill. As he did for several Newmont operations, Frank W. McQuiston carried the responsibility for overseeing the design of the Carlin mill. Using a conventional Merrill-Crowe precipitation process, the mill was designed and built by the Bechtel Corporation of San Francisco. An unusual feature of the mill is its extensive use of automation in control. Three men run the entire operation from two control centers. One of these centers controls truck movements and the crushing plant. The other center controls the mill operation from grinding to gold recovery. The second center also monitors operation of water wells, water reclamation from the tailings pond, and power plant operation. Computer operation is a part of this control system. With a capacity of 2,400 tons of ore daily, the mill initially achieved recoveries of around 92 per cent, but as more and more of the carbonaceous gold ore turns up in the mill, the recoveries have been running lower and are now around 88 to 90 per cent.

Carlin's output in the form of 1,000-ounce bars, worth from $35,000 to $42,000 each, was sold until March, 1968, to the U.S. Mint in San Francisco. At that time, the U.S. Treasury stopped buying gold, and Carlin's output now goes to custom refineries, which sell it on free gold markets.

Since 1965, Carlin has turned out about $50 million worth of gold. In annual production of gold, Carlin is surpassed in the United States only by the Homestake mine at Lead, South Dakota, and by-product gold recovery from Kennecott's Utah Copper mine at Bingham Canyon, Utah. From the very beginning, Carlin has been most rewarding, in satisfaction to the men who planned it, and in profitability to the parent company. Three years after production of its first gold bar, Carlin had paid off the original investment, and in the interim, had contributed about $3 million in dividends to Newmont—quite an achievement when you consider that in recent years gold mining has been regarded as a decreasingly profitable activity.

From the point of view of exploration, Carlin is quite significant in several ways. Had Newmont's geologists been looking for "anything," they might never have found Carlin's gold, for none of it is visible either on the surface or in the mine. The time-honored approaches to gold discovery would not work on a deposit like Carlin, for there are no adjoining placer deposits to call attention to the orebody. Nor is there any float to be followed up to the orebody, as prospectors have always done. Furthermore, Carlin is one of the very few successful open-pit gold mines in the world. In short, Carlin has turned out to be exactly what Bob Fulton and his associates were looking for, a large and productive gold property susceptible of low-cost mining.

Here, again, is an example of a balanced combination of exploration capability, metallurgical skill, and financial acumen, all combined with the essential element of good judgment. Malozemoff might have been speaking of Carlin when he once said, "As in all of our investments, we evaluate the causes of observed phenomena, test each variable, and define it to eliminate it from the spectrum of the unknown. We proceed step by step, not spending very much at each step, making a judgment on the basis of a mass of engineering criteria and detail." This philosophy has guided Newmont until today, and it can be relied upon to guide the company tomorrow.

Offshore oil, Louisiana

15 Newmont in the Petroleum Industry

Does a nonferrous metal mining company really belong in the oil industry? Debate on this question has surged around many a luncheon or board-room table, but there seems to be no definitive answer. Even Newmont, which has been in the oil industry almost throughout its life, would have to supply a somewhat mixed answer, depending on who in the organization had been asked the question.

Colonel Thompson put Newmont into the petroleum industry in about 1925. Thereafter, to remain, or not to remain, in the oil industry has been a point at times hotly debated within the company, yet on the whole, Newmont has done rather well on its ventures into petroleum. However, this has not been accomplished without a certain amount of internal stress, and it has been done only by recognizing certain basic differences between nonferrous metal mining and petroleum industry operations.

Aside from investing in a bewildering variety of oil stocks of widely varying profitability, Colonel Thompson left no record of attempting to find and to produce oil himself until the early 1920's. Hearing a rumor of a new discovery of oil in Utah at that time, Thompson promptly sent Henry Krumb to look it over, thereby demonstrating a sublime faith in Krumb's ability to diagnose a potential oilfield as well as a porphyry copper. Nothing daunted, Krumb examined the structure at the discovery site, and because of the domelike nature of the formation, evident on the surface, he concluded that it at least had a possibility of holding oil. He could not bring himself to recommend it, however, and although Thompson had, in the meantime, worked up a good deal of enthusiasm for Utah oil, nothing really came of it.

In a rare bit of peevishness, Thompson is said to have told Krumb after this episode, "You have ruined more schemes of mine than any man living." This must have amused Krumb immensely, because he and Thompson both knew that Thompson was almost completely dependent on Krumb's judgment of mineral properties, particularly after the Mason Valley experience described in Chapter 1.

Thompson and Newmont did not again move to become actively engaged in petroleum production until about 1925, when Thompson became interested in the Orange oilfield in Texas, where some of his friends, Bernard Baruch among them, were drilling what they thought would turn out to be a producing field. As frequently happened in those days, there appeared to be undue confusion as to ownership, and Thompson sent Judge Ayer down to try to straighten things out. Ayer's clear, cool, legal mind was extremely useful here, and before long, the conflicting claims were resolved and the several interests consolidated under the name of Salt Creek Producers, and the company began to acquire options on several surrounding properties. Before the Salt Creek Producers company could get into production. Thompson became acquainted with a company that looked better to him. Continental Oil Company of Maine had been formed some years previously to take over the Standard Oil properties in the Rocky Mountain states that had become available when the U.S. Department of Justice, in implementing the antitrust laws, had enforced the dismemberment of Standard Oil. Liking what he saw of Continental of Maine, Thompson negotiated to swap his interest in Salt Creek Producers in return for an interest in Continental of Maine. Newmont was represented on the Continental board by Fred Searls, who had joined the company in 1925.

At this time, J. P. Morgan & Co. were acting as bankers for the Marlin Oil Company, a small but growing company with headquarters in Ponca City, Oklahoma. The Morgan partners thought that Marlin was being run somewhat extravagantly, with more emphasis on polo than on prospecting. Dan Moran, a hard-driving man from the Texas Company, was sent in to get Marlin back on the track of finding and producing oil. The Morgan Bank was also the agency instrumental in bringing about a merger that took place in early 1929 between Marlin Oil Company and Continental of Maine. This brought Fred Searls into direct contact with Dan Moran, and the two aggressive personalities struck sparks. Following an argument over certain properties in West Texas, Searls resigned and Franz Schneider took his place on the Continental board. This happened in 1930 after the merged companies were incorporated in Delaware as Continental Oil Company. Newmont's interest in Continental has varied from time to time, but for the past ten years, Newmont has been the largest single shareholder in Continental Oil Company. Schneider remained on the board for forty-one years, retiring in May, 1971. Newmont is now represented on the Continental board by Plato Malozemoff.

Still on the lookout for suitable oil investments, Newmont, at the instigation of Philip Kraft and Franz Schneider, bought, in 1932, a large block of stock in Amerada Oil Company, in which Henry Krumb was already a shareholder. Krumb hung on to his Amerada shares, and after his death in 1958, his Amerada stock, for which he had paid $8,500, was worth over $500,000. Newmont, on the other hand, sold its Amerada in 1945, a move that Fred Searls years later characterized as "the greatest mistake made during my administration of Newmont's affairs."

Although clearly Newmont's well-being during the Depression was based almost entirely on gold mining, all through the 1930s the petroleum industry came to look more and more attractive to Searls. He was well aware, too, that in Phil Kraft, Newmont had a staff member who was quite familiar with the petroleum industry; Kraft had spent six years with the Bishop Oil Company in California before joining Newmont.

With Searls's encouragement, Kraft developed a more active interest in the oil industry, with the result that during the mid-1930's, Kraft entered into agreements with Amerada and Continental for financing exploration in selected areas of Louisiana, Texas, and the entire Rocky Mountain area. As this activity grew, a separate company became necessary, and on October 15, 1938, Alder Oil Company was incorporated with Phil Kraft as president, Franz Schneider as chairman, William T. Smith as treasurer, and later on, Roy Bonebrake as vice president. The name of this wholly owned Newmont subsidiary was derived from Alder Gulch, Colonel Thompson's birthplace in Montana.

One of Alder Oil's first acts was to open an office in Houston with Lou Roberts, a geologist, in charge. Alder Oil Company at first made no attempt to discover or to produce oil on its own, but was content to buy mineral and royalty interests that were strategically situated on trends or in areas that were considered of potential value for future production. Although it did not actively seek producing royalties, probably because of the higher first cost, Alder Oil began to do quite well through royalties acquired in the Vielle Platte oilfield in Louisiana, which had been discovered in 1937.

In an aggressive purchase program directed by Phil Kraft and Lou Roberts, Alder Oil put together over 13,000 acres in Louisiana, Oklahoma, Texas, Kansas, Mississippi, and Illinois. By 1940, four wells were producing income for Alder, and the company had acquired thirty-four productive royalties in all. Profits from oil for that year totaled about $60,000, not of great significance to the parent company, except that it indicated a profit could be made in the oil industry by a metal mining company.

Fred Searls became about as fascinated with the petroleum industry as he was in his operations in the "Wall Street Stope." In May, 1957, he wrote for company consideration a lengthy memo in which he compared the growth of the petroleum and the copper industries over a forty-year period. These figures, of course, showed a much greater growth for petroleum than for copper, but the bare figures by no means told the whole story of the tremendous financial outlay required in the petroleum industry and the high rate of risk involved. Nevertheless, Fred said in his memo that "while Newmont has made progress in both industries—oil and copper—I think it is fairly reasonable to assume that if we had put the same money and effort into the oil business as we have addressed to copper, we'd have had at this time a much larger net worth." Knowing something of Fred and his associates at that time, one can easily imagine that the resulting discussion was lengthy and possibly stormy. There is no evidence whatever that the course of the company was altered by Fred's opinion of that moment.

Alder Oil Company proceeded with reasonable success for several years, but then in 1945, a succession of dry holes resulted in a loss for the year of $124,000, a somewhat discouraging development, but not enough to alienate Fred Searls. In October, 1944, Newmont had incorporated a second wholly owned subsidiary called Newmont Oil Company, for the purpose of acquiring producing royalties, and in 1946, Newmont Oil was operating profitably. That year the parent company decided to secure the benefits of consolidation by merging the two companies, and on December 31, 1946, Alder was merged into Newmont Oil Company, which was thereafter directed by the same officers.

The merger with Alder brought into Newmont Oil Company the services of Lou Roberts and another experienced petroleum geologist, Oscar Champion, who had joined Alder in June, 1944. Champion had seen service with Seaboard Oil and the

Amerada Oil Company. Champion and Roberts established an office in Fort Worth, which later became the field office of Newmont Oil Company, at which time Champion became a vice president of Newmont Oil.

The New York office of Newmont Oil Company, under Phil Kraft, handled all land, legal, and accounting records of the company until 1962. Kraft continued as president of Newmont Oil until 1955, when he retired officially, but continued as a most active chairman of the board until his death on December 2, 1968.

Not many companies, even today, can boast of having a woman as an officer, yet Newmont Oil Company felt such feminine influence as long ago as 1945, long before today's Women's Liberation Movement. Alice Langlois was a vice president of Newmont Oil Company from 1946 to 1954. A native of Canada, fluent in French, and a graduate in geology of Columbia University, Alice had joined Alder Oil Company shortly before World War II. She transferred from Alder Oil to Newmont Oil in 1945, becoming assistant secretary and assistant treasurer, and, in February, 1946, she was elected vice president of Newmont Oil Company. A competent and attractive woman, as vice president Alice Langlois ran the office and participated in, and kept track of, all the acquisitions of mineral and royalty interests of Newmont Oil during her tenure. When Newmont Oil Company moved its office to Houston in September, 1954, Alice Langlois chose to resign as vice president and to move abroad with the Mobil Oil Corporation in order to live in Paris, which she had always wanted to do.

From 1955 until 1962, Plato Malozemoff was president of Newmont Oil and remained on the board until 1965. Dr. Robert S. Moehlman became president of the company in 1962, after resigning as executive vice president of Austral Oil Company, whose Houston office he had inaugurated in January, 1951. Moehlman had worked briefly for Newmont in the summer of 1935, as a temporary employee doing geological work for Idarado in Colorado. His first assignment was to spend two weeks in

Paris, France, in March, 1962, helping Newmont evaluate its position in the newly discovered Rhourde el Baguel oilfield in Algeria, the development that Phil Kraft always thought of as his greatest accomplishment.

At present, Dr. Moehlman is president of Newmont Oil, Jack E. Thompson, a director and executive vice president of Newmont Mining Corporation, is chairman of the board, and Jesse L. George, Jr., and E. Duwain Whitis, are vice presidents. George E. Zubrod, Jr., is controller; Frank Johnson is land manager; Edward I. Barton is foreign exploration manager.

In general, the activities of Newmont Oil Company may be divided into three major categories: the onshore royalty and mineral interests; the offshore properties in Louisiana, and the waterflood operations in New Mexico.

The first of these three categories, purchase of onshore royalty and mineral interests, was the door through which Newmont entered the petroleum industry. Mindful that his company would be both unwilling and unable to assume the risk of an all-out wildcat exploration program, Phil Kraft cautiously set about acquiring mineral rights and royalty interests in producing fields or in areas that were considered potentially productive. In addition to Oscar Champion's assistance, Phil was aided by L. P. Teas, a geologist in Houston. Kraft also acquired a substantial number of royalties in Texas and Louisiana through the help of Julius Fohs, formerly a geologist for the U.S. Geological Survey. In the four years following its incorporation in 1944, Newmont Oil invested about $1.2 million in buying mineral and royalty interests. So carefully selected were these interests, that Newmont Oil's activity in this area has been in the black every year since 1949. During that time, gross royalty income has ranged annually between $900,000 and $1.5 million. In the twenty-three-year period from 1949 through 1971, the onshore royalties and mineral leases show a cumulative net cash flow of $20.2 million. Between 1949 and 1956, Newmont Oil spent about $3 million acquiring royalties and mineral leases. Since 1956, expenditure on this source of income has averaged under

$50,000 per year, yet the cash throwoff is still running about
$1.3 million per year, testimony to the fact that Newmont Oil
chose wisely in buying extensive royalty and mineral interests so
that new wells have been brought in to replace declining production from older wells.

Having achieved a solid, if not spectacular, success in the
first four years of its existence, Newmont Oil Company in 1949
was ripe for a suggestion made to it by L. P. Teas. In his work
as a consulting geologist, Teas had learned that Magnolia Petroleum Company, a wholly owned subsidiary of Socony-Vacuum
Oil Company, Inc., which in 1966 changed its name to Mobil Oil
Corporation, had an offshore oil project in hand that was really
too much for the company to handle alone. One of the pioneers
in offshore drilling, Magnolia had acquired from the State of
Louisiana about sixty-four leases covering approximately 220,000
acres in an area reaching between 10 and 30 miles from the shore-
line in water between 10 and 50 feet deep. By 1949, Magnolia
had sunk a good deal of money into the offshore project and had
drilled a number of dry holes, but only three productive wells.
Furthermore, even the productive wells were gas producers,
which at that time, was not salable from offshore sources.

To finance further exploration and development, Magnolia
thought it wise to invite partners, and when Phil Kraft heard
about it from Teas, he became quite enthusiastic. Kraft began
negotiations with Magnolia, and at the same time, tried to
arouse interest in such a venture among his associates, Fred
Searls, Franz Schneider, and Roy Bonebrake.

Phil Kraft's discussions with Magnolia developed rapidly to
the point where Magnolia offered Newmont a half interest in
the offshore venture for $18 million, much more than Newmont
Oil wished to take on at that time in a single venture. However,
the idea was attractive, and to maintain an interest in it, and to
bring in a substantial partner, with whom it had a community
of interest, Newmont invited Continental Oil to take up three-
quarters of the half interest offered by Magnolia to Newmont.

Continental accepted the offer, and on October 1, 1949, an

agreement was executed among Magnolia Petroleum, Continental Oil and Newmont Oil, which included the following three elements:

1. Continental-Newmont paid Magnolia the sum of $6,428,500 as consideration for an initial minimum 10/28 interest in the leases, equipment, and other facilities.

2. Magnolia-Continental-Newmont agreed to spend an additional $10 million on development and exploration, of which Magnolia would pay 18/28.

3. Continental-Newmont had the option, after the initial $10 million program, to increase their aggregate interest to one-half by making installment payments totaling $4,488,200.

The agreement was fully carried out by Continental and Newmont, so that the interests of the three companies became fixed at Magnolia, one-half; Continental Oil Company, three eighths; Newmont Oil Company, one eighth.

This MCN venture was prosecuted as a massive continuing investment, with Newmont's share of exploration and development costs climbing from $681,000 in 1950 to a peak of $4.1 million in 1957, thereafter continuing at about $1 million per year.

Because the MCN venture was a pioneer in offshore oil exploration, the group necessarily had to bear some costs of experimental design and construction. Most significant of these costs were the changes made necessary by revision and improvement in design and construction of the offshore drilling and production platforms. Protection from the effects of wave impact was one strong consideration, but protection against damage through accidental blowout of a well, or by fire, has always been a primary consideration of the MCN venture. For example, during 1970 and 1971, the MCN group spent approximately $7.5 million on safety equipment designed to prevent disasters, such as the fire and massive oil spillage that occurred in 1970 at one of the other drilling locations in the offshore Louisiana area. Despite these high development costs, the MCN venture has been remarkably successful. Newmont's share of production in MCN has climbed from a token $19,000 in 1950 to $6 million in 1971.

To say simply that the MCN venture was successful, is to ignore the great sums of money that must be poured back into exploration, development, and equipment in order to keep such a property as this in production. For example, in the twenty-three years of its existence through 1971, the MCN group produced over $475 million in oil and gas. The cumulative operating income after deducting production tax, refund on gas sales, and production costs, totaled $337.6 million. In the same twenty-three-year period, the MCN group spent in capital costs for exploration, development, equipment, and leases a total of $244 million, leaving a net cash flow of $93.6 million. Newmont Oil's share of this was $11.7 million. However, much of the heavy expenditure is behind the venture, and the Louisiana offshore venture should continue to be reasonably profitable.

By 1956, with the MCN venture well established, but with payout still a long way in the future, Newmont Oil Company was looking about for an investment with a quicker return. Oscar Champion then recommended buying waterflood properties, and the company began a search for such opportunities. In 1958, two waterflood properties were acquired, a small one in West Texas and interests in a larger area in Loco Hills about 22 miles east of Artesia, New Mexico.

In the petroleum industry, the term "waterflood" refers to the practice of injecting water into the ground under pressure in a previously producing oil reservoir. Certain of the old wells can be converted to injection wells in a regular pattern so that oil remaining in the formation will be pushed toward producing wells. In a typical waterflood operation, five to ten barrels of water will be injected for each barrel of oil produced. Such secondary operation may recover an amount of oil equal to that produced in the primary producing period. In a field handled in this way, the reservoir should be operated as a whole in order to allow the efficient injection of the water, movement of the fluid underground, and the recapture of oil in producing wells regardless of property lines. This may require cooperation among many parties owning interests or operating in the area through agreement on sharing among the various interests represented.

Response was rapid to the initial Loco Hills flooding operation, so much so that following an investment of $1.4 million by the end of 1959, payout had been achieved by the middle of 1961. By the end of 1969, the Loco Hills waterflood area had a cumulative cash flow of $3.25 million.

Despite its lack of glamour, waterflooding has been a profitable item for Newmont Oil Company, yielding an operating income in 1971 of $741,000. From the beginning in a part of the Loco Hills area in 1958, Newmont Oil has expanded its waterflood activities into two other areas in New Mexico and into a nearby area in Texas as well. All New Mexico waterflood operations are supervised from a district office in Artesia, where Charles Joy is superintendent. A district office has also been established at Snyder, Texas, under John Fagin, to supervise waterflooding at Sharon Ridge where the Ira Unit became effective in 1968. As a part of waterflood production, Newmont Oil had also to establish the Yucca Water Company to locate and to supply water for the Loco Hills operations.

All of the foregoing activities of Newmont Oil Company, while indicating growth, are certainly not the stuff of which huge corporations and oil millionaires are made. One suspects that a thought such as this must have been in Phil Kraft's mind when, shortly after the MCN venture was established, he vigorously advocated what amounted to a wildcatting exploration program for whole new oilfields. At this point, the difference between mining companies in the oil business and petroleum companies in the oil business had to be recognized and re-established. Searls, and later Malozemoff, had to restrain Kraft, because they knew that Newmont as a mining company simply did not have the resources available to take the risks that are commonplace in the petroleum field. A string of twenty dry wells will not discourage a well-financed petroleum company, because the twenty-fifth or the thirtieth may well make up for all the others. Newmont could not take such a risk, simply because it could not afford to run out a string of fifteen or twenty dry holes with no resulting income. Therefore, Malozemoff was always inclined to hold wildcatting down, and he urged the Newmont Oil Company management to concentrate on land plays and a search for participation in potentially large and promising ventures, such as the MCN offshore venture.

Certainly this is a prudent, conservative policy, but a difficult one to maintain, especially in the petroleum industry. In 1958, Newmont Oil came across an opportunity to secure a partnership in an offshore drilling program in California. A syndicate was being formed between the Texas Company and Monterey Oil Company, with a 25 per cent interest being offered to Newmont Oil. Clearly, the risk involved was substantially greater than any such previous investment by Newmont Oil, but if successful, it was thought that the venture could have returned a profit to Newmont Oil well beyond anything thus far experienced. Phil Kraft was in favor of the venture and wanted to finance the Newmont Oil interest by bank borrowing. Although Malozemoff was intrigued by the rich potential payout, he considered the proposition too risky to involve other people's money. Accordingly, the entrance of Newmont Oil into the partnership was financed by equity.

The drilling proceeded, and although some commercial production possibilities developed, there was not enough oil revealed even to pay back the purchase price. In October, 1961, Newmont Oil sold its 25 per cent interest at a loss of about $5,900,000, a loss that was borne by the parent company. After-tax write-offs did bring the actual loss to below $3 million. Newmont's banker friends were aware of Malozemoff's reluctance to use borrowed funds for this venture, and rather than damaging Newmont's reputation, the use of equity financing actually enhanced Newmont's credit standing.

"The California offshore oil deal was one of my biggest mistakes," says Malozemoff. "It was like playing the numbers in roulette. The odds were too high, but I got carried away by the thought of what we'd win if we were lucky. Not that it all depended on luck, but the gamble was just too great for a mining company to risk that much on one shot. A big oil company can do it, because it would have several such deals going, but we're not a big oil company. At least, not yet."

At one time or another, probably everyone in the petroleum business has become interested, or involved, in Venezuelan oil. In 1957, Newmont Oil joined a syndicate to prospect an area in Lake Maracaibo at a cost of about $49 million. The syndicate acquired a lease and began drilling. The partnership included San Jacinto Petroleum, Phillips Petroleum Company, with about 75 per cent ownership, and several other companies, one of which was Newmont Venezolano Limited, formed especially for the purpose.

Newmont Oil contributed $2,400,000 to the syndicate. With no further contribution, Newmont was to receive, on a sliding scale, from 18 per cent of the proceeds of the venture down to 10 per cent, until it had received seven times the original $2.4 million, receiving 6.667 per cent thereafter. This is mentioned only as an illustration of the kind of deal involved in a project of this sort, because none of this division of profits ever took place. Although the first hole indicated a small production of oil, subsequent drilling was most discouraging, and in 1960, Newmont withdrew from the syndicate and wrote off the original investment.

With the offshore California and the Venezuelan disappointments in back of him, Phil Kraft was particularly pleased when, in Algeria, Newmont at last participated in the discovery of a large and highly profitable new oilfield. On Phil Kraft's recommendation, Newmont, through a new company, the Newmont Overseas Petroleum Company, did participate in what turned out to be a major oilfield in the Algerian Sahara. It was a source of great disappointment to Kraft that this venture for which he had such high hopes, was frustrated by Algerian nationalistic developments.

Early in 1957, Kraft, who had developed a wide acquaintance in France, and particularly in Paris, was abroad investigating an offer from a French company (SAFREP), to which he was introduced by Lazard Frères, to participate in an onshore oil venture in Gabon in West Africa, which he wisely turned down. However, on this trip he learned of discovery of oil in Algeria in late 1956, and on his return he recommended that Newmont negotiate a participation in a promising venture in the area. Malozemoff and Jacques Leroy, Newmont's Belgian-born attorney, met in Paris and laid the groundwork for the first entrance of American capital into Algeria. Eventually, this took the form of a consortium of five companies, three French and two American.

Among other areas, the consortium explored about 3,900 square miles of the Sahara Desert in Algeria. The venture was re-markably successful; seven producing wells were brought in at a point known as Rhourde el Baguel. Production of this new field was limited to about 21,300 barrels per day for lack of an ade-quate pipeline to the coast. Construction began on a new com-mon carrier pipeline, which could allow the new field to de-velop into one of the industry's great producing areas, and production eventually soared to 80,000 and then 100,000 barrels per day.

However, from the beginning, Newmont was uneasy about marketing its share of production from the field, never having developed marketing facilities for oil. Dr. Moehlman initiated negotiations with Tidewater Oil Company in 1962, which led to the conclusion of an agreement early the following year whereby Tidewater and SAFREP acquired half of Newmont's interest in the Rhourde el Baguel property on the condition that Newmont's share in the profits would be proportionate, to its interest, but that Newmont need not put up additional funds for the develop-ment of the oilfield, and that Newmont would also receive $1.2 million in payment of one-half of its prior expenditures. Within about two years, Newmont had its investment back and was earning a profit.

Then, in 1967, the Israeli-Arab War broke out, and follow-ing it, all profits and assets of American companies were frozen in Algeria. Newmont Overseas Petroleum Company's bank bal-ances and other assets thus were blocked within the country from

1967 to 1971, by which time they aggregated over $4 million. Following lengthy and repeated efforts to secure release of these funds by negotiation with the Algerian government, Newmont Overseas Petroleum Company initiated arbitration proceedings in Paris in 1970 to attempt to recover its blocked assets. In November, 1970, the Algerian government nationalized all the properties of Newmont Overseas Petroleum Company, and no income has been booked thereafter. Inasmuch as no funds had been transmitted from Algeria to the parent company since June, 1967, this seizure had no new effect on the earnings of Newmont Mining Corporation. Early in 1971, negotiations with the Algerian government, carried on for Newmont largely by Jacques Leroy, suddenly made progress, resulting in settlement of all Newmont's claims against the Algerian government. The compensation was a fraction of the value of the oil and mineral holdings, but the company had no alternative but to accept.

More recently, Newmont Oil became involved in the exciting petroleum development on the Alaskan North Slope. The company was able to obtain an 18.5 per cent interest in prospective leases covering 20,453 acres in the Prudhoe Bay area and Cook Inlet. At present, largely owing to disagreement on how petroleum production from the North Slope could, or should, be brought back to civilization, the future of this new field is somewhat uncertain. Furthermore, conflict between the production-minded and the environmentalists is further complicating the development.

During 1971, Newmont Oil Company spent a total of $2,300,000 for exploration, acquisition, and development of new properties. Gross sales of Newmont Oil in 1971 were at an all time high of $10,319,000. Net income for 1971 rose to $2,009,000 from $1,501,000 in 1970.

All in all, Newmont, despite reverses in California, Venezuela, and Algeria, has not done badly in the petroleum industry. Although it was unspectacular, much of the company's success had been due to the careful selection and development of producing properties initiated by Phil Kraft, aided by L. P.

Teas, and carried on by Oscar Champion, and, currently, by Dr. Moehlman. However, an even larger factor has been the financial planning and the sound backing provided by Newmont Mining Corporation. In the beginning, while Newmont Oil was going through the long period of planned acquisition and the heavy expenditure required to develop the offshore Louisiana venture, Newmont Mining Corporation supported the oil company through loans and advances. Total investment by Newmont in the oil company as of December 31, 1962, was $19.6 million before any tax credits. During this time, the tax losses of the oil company had been used by Newmont in its consolidated tax statements, and the resulting deduction brought Newmont's after-tax investment in Newmont Oil down to about $9.5 million.

The date of December 31, 1962, is mentioned because at that time Newmont Mining Corporation instituted a change in its method of handling Newmont Oil Company that had been advocated for a long time by Walter P. Schmid, who was then controller of Newmont. It was Schmid's contention that Newmont Oil should be restructured, and that the various loans and advances made by Newmont to the oil company should be converted to represent contributions of capital. This would enable Newmont Oil to operate in the black, and to begin paying tax-free dividends to the parent company as a consolidated subsidiary.

This change was made at the end of 1962, and the Newmont investment of $19.6 million was duly entered as a contribution to the capital of Newmont Oil. Accordingly, in 1963, Newmont Oil paid Newmont Mining its first dividend of about $600,000. Through 1971, Newmont Oil had paid tax-free dividends to Newmont of about $10 million. It should be noted that there was no loss carryforward to be credited against Newmont Oil Company's taxable income in 1962 and later years, because the deficits from 1944 through 1961 had been properly used by Newmont Mining in its consolidated returns.

Newmont's net investment in the oil company was increased in 1969 by $7.8 million, which was used to purchase the interest in the acreage on the Alaskan North Slope. This brought Newmont Mining's total net investment after taxes to $17.4 million, against which are balanced the dividends paid to Newmont by the oil company from 1963 through 1971 of $10.1 million. Furthermore, Newmont Oil now has no outstanding debt and, in addition to assets in the ground, had net working capital of $2.3 million at the end of 1971.

In terms of return on investment, the operations of Newmont Oil Company do not stand comparison with such Newmont investments as that in O'okiep, Tsumeb, or Magma. However, it now appears that the long and expensive period of painstaking development is beginning to pay off, and the future is a bright one. To the question of whether or not a metal mining company belongs in the petroleum industry, Newmont can respond with a qualified affirmative.

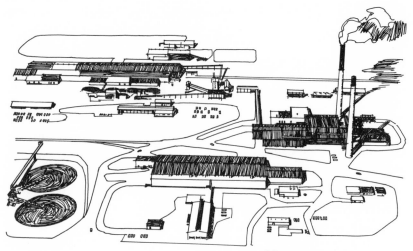

San Manuel's plant today

16 Newmont — Today and Tomorrow

Anyone who has attempted to write history will have discovered that it is much easier to discuss and to analyze events of a hundred years ago than it is to record and to interpret what is taking place today. For one thing, there is no one left around who can quarrel with your interpretation on the ground of actual experience. For another, the events even of fifty years ago are all safely tucked away in perspective, and the results of each action have long since been recorded as good or evil.

Thus in touching on Newmont today, one has the disquieting knowledge that whatever one writes today may turn out to be a misinterpretation tomorrow. It happens that Newmont Mining Corporation is today spinning out history quite as fast as it ever has, and the printed word must necessarily be left hopelessly behind.

For example, Magma's plans for control of air pollution at the San Manuel smelter have already been discussed, but at the time of writing, Magma had not yet fully resolved the problems of how to utilize the full capacity of the smelter (1 million tons of concentrate annually) without undue discharge of sulfur dioxide, or of how to dispose of the smelter's production of sulfuric acid. Neutralization of the acid would be expensive and a dead loss. Using the acid to leach oxide ores is obviously preferable, but such ores are not readily available, as those nearby are owned by others who naturally want to take advantage of a "buyers' " market in the light of large surplus production of acid by other smelters as well.

Elsewhere under exploration or study are other copper orebodies in Arizona, a nickel orebody in Alaska, base metal prospects in Africa, lateritic nickel deposits in Indonesia, prospective oil properties in the North Sea and Alaska, as well as investigation of new and improved metallurgical processes. Some of these may furnish significant material to begin the history of Newmont's next fifty years.

A good part of this forthcoming history, at least that of the next few years, will be devoted to Newmont's methods of dealing with environmental problems. In addition to air-pollution control by Magma, Newmont is attempting to forestall pollution problems before they develop.

Disposal of mill tailings has always been a problem of concern to mining companies, and Newmont for many years has been attempting rather successfully to prevent contamination of air by dust not only from tailings deposited by mills which it owns or operates, but from nearby tailings left by former operators as well. Idarado is carrying on such a program of seeding old tailings deposits for grass coverage. In fact, no Newmont-managed property contributes to watershed pollution in its area.

Turning now to other new developments, among Newmont-managed-or-owned mineral prospects, the one nearest to commercial development is the Vekol Hills copper property on the Papago Indian Reservation in south central Arizona. Neither extremely large nor high grade, the Vekol orebody presents the same problem that faces the developer of any new sulfide copper orebody in the United States or Canada. This is the simple one of what to do with the sulfide copper concentrates, inasmuch as there is no smelter anywhere in the world that, in 1972, was anxious, or even able to increase its intake of this material. Any such smelter, including San Manuel, that might purchase the Vekol concentrates must consider the added cost of eliminating sulfur from the gases produced by smelting this additional tonnage. This extra cost would involve an extra burden for Vekol and will require a long and careful prior look at the economics of the entire operation.

In Canada, Granduc Operating Company has shown improved results, after a halting and disappointing start in 1971. Granduc continues the effort to increase the grade of the ore being mined, as well as to increase the daily mine production and reduce costs. At Similkameen, construction is complete, and this 15,000-ton-per-day-open-pit copper mine entered initial production in late March, 1972.

In Australia, Newmont Proprietary Limited continues under the direction of Robert J. Searls, the son of Fred Searls, Jr. Newmont's interest in Australian mining developed a good many years ago in conversation among Fred Searls, John Gustafson, and L. B. Robinson, the son of W. S. Robinson, who was known throughout the world as the most prominent member of the Australian mining industry, particularly lead-zinc mining. Even earlier than these conversations, Fred Searls's globe-girdling ambitions had caused him to send Joe Thorne, his exploration partner, to Australia in the late 1920's or early 1930's, but nothing tangible resulted from this trip. Thorne stayed in Australia and became well known and loved in the Western Australia gold fields.

In 1947, there was established among Zinc Corporation (an Australian company), St. Joseph Lead Company, and Newmont, a joint venture to look for new Australian mining opportunities. For two years, this joint venture conducted explorations in or near old mining camps in central and eastern Australia, under the general direction for Newmont of P. C. Benedict, and managed by Sir Maurice Mawby, of Zinc Corporation, who later became managing director and then chairman of Conzinc Rio Tinto of Australia. Although a good many promising prospects turned up, there seemed to be nothing large enough to attract an investment at that time. Furthermore, metal prices in 1948 were extremely low.

Newmont made further exploration attempts in Australia in the 1950's, but not until 1964 was a fairly strong exploration campaign developed. Bob Searls visited Australia that year and got Hal Norman to accompany him on part of the trip. They spent most of their time going over the pre-Cambrian shield in Western Australia. On this trip, Searls and Norman found strong similarities between geologic structures in Western Australia and the pre-Cambrian shield in Canada and also came to the conclusion that standard methods of prospecting were not adequate for the problems presented there.

In 1966, Newmont Proprietary Limited, was incorporated in Australia with offices in Melbourne, Bob Searls becoming managing director. Jacques Claveau came down from Spain to assist in the Australian exploration activity, and Claveau and Searls, together with Bruce Webb, a well-known Australian geologist, were active in Newmont Proprietary thereafter. However, in September, 1972, Bruce Webb left Newmont to assume the important post of Director of Mines for South Australia.

In general, Bob Searls and his Australian company prefer to work with partners. For example, in the Norseman area, Newmont was looking for nickel together with Norseman Gold Mines N.L. The company is also looking for uranium deposits in the northern territories. A number of other gold, copper, lead, zinc, silver, and nickel prospects are under exploration, but none considered of commercial importance has yet been found.

A large area of Indonesia is under investigation for copper by Newmont Proprietary Limited with several partners. Exploration for oil in the Bonaparte Gulf west of Darwin started some years ago, and is continuing. Newmont has a 15 per cent interest in an Indonesian company, P. T. Pacific Nikkel Indonesia, which is developing laterite nickel deposits in Irian Barat. Over 330 million tons of ore averaging 1.44 per cent nickel is now in sight on Gag Island and other islands nearby. Such a reserve has promise of a profitable operation that may take three to four years and several hundred million dollars to inaugurate, however.

This thumbnail sketch of what Newmont is doing today would be incomplete without reference to the men doing the job. A listing of the men now active in Newmont appears in an appendix. Many of these men have appeared briefly thus far in this narrative, but there are new faces among the management. Plato Malozemoff, president and chairman, no longer has the direct assistance of some of the former Newmont stalwarts, although Franz Schneider, at 85, is still in regular attendance, and still available for consultation.

Among the older members, Newmont has lost by death Fred Searls, in October, 1968; Phil Kraft, in December, 1968; Henry DeWitt Smith, in October, 1962, and Carroll Searls, in November, 1970. The activities of these men created the Newmont history in the earlier years, and they have been difficult to replace.

In particular, Fred Searls was certainly as widely known and highly respected as any individual in the world's mining industry. If one achieved friendship with Fred Searls, he at once acquired an introduction to friendship with hundreds of men in the industry in countries wherever mining was carried on. Fred had this ability to attract widespread friendship to an extraordinary degree, which is the more remarkable in that he was never at pains to hide his lack of interest in a man or a situation of which he did not approve.

Somewhat a creature of habit, Searls at first flatly refused to move his office from 14 Wall Street when Newmont, in October, 1955, moved its headquarters uptown to 300 Park Avenue. After everyone had left, Searls sat at his desk at 14 Wall, obdurate but lonely, until the silence got too much for him, whereupon, some days later, he showed up at 300 Park, with only the gruff question to Malozemoff, "Where do you want me to sit?"

Moving the office uptown was one of three Malozemoff-sponsored actions after he became president in 1954 that Fred Searls strongly opposed. The third, chronologically, was Atlantic Cement, to which Searls never became reconciled, and the second was a stock option plan for key employees that came closer than any other issue to causing a serious split between Searls and Malozemoff.

Shortly after he became president, Malozemoff engaged a firm of management consultants to study the Newmont set-up and recommend improvements. Searls and some of the other old-timers didn't think much of the idea, but Searls, at least, said nothing while the study was going on. After all, Searls had told Malozemoff that he was not going to interfere in Newmont's affairs after Plato became president, and rather than fight over something of which he disapproved, he (Searls) would resign from the board if he couldn't accept it.

The consultant's report called for, among other things, an incentive plan based on stock options to a few key employees. Malozemoff showed it to Fred, who objected somewhat to the options, because certain old employees were to be left out. However, he did not express his dislike to Malozemoff, but went instead to Margaret Biddle, W. B. Thompson's daughter, in Paris.

On her next trip to New York, Mrs. Biddle had lunch with Plato alone, and the conversation went something like this:

"What are you and Fred arguing about? He is so upset he gets almost incoherent when he talks about it. I certainly want to hear your side of it."

"I didn't know we were arguing," said Plato, feeling as if the ground had rocked under him, "I know Fred doesn't like some of the things I'm thinking of doing, like a stock option plan we have in mind. Could that be it?"

"Yes, indeed," said Mrs. Biddle, "He hated having that company come in and poke around the office, and he hates that stock option plan even more. Didn't he tell you? Well, then, I'll tell you."

Mrs. Biddle then went on to explain that Fred couldn't bear

to see stock options, or other bonuses, granted to only a few employees, however important they might be. He was quite distressed at seeing some older employees about to be overlooked, and despite his previous assurances to the contrary, he might even fight openly for his point of view. Mrs. Biddle then proposed a compromise plan to bring in more employees, to which Malozemoff made a counter proposal, and in short order the issue was resolved between them. Fred accepted the compromise, and never again mentioned the point. Mrs. Biddle, after this statesmanlike effort, proved to be quite helpful to Malozemoff and gave him her complete confidence and support until her death of a stroke in 1956.

Actually, there was no more conflict between Searls and Malozemoff than between any other two men in Newmont. Both men had strong opinions and disagreements often occurred, yet, as among all the Newmont men, really serious quarrels never developed.

Malozemoff learned restraint in his early years under Fred, and the value of the advice and assistance of associates. As a young man, Malozemoff fixed on the rule of reason and logic as the inflexible guide for his life, and to some extent he has been battling with reality and its frequent illogical developments ever since.

That he has won more such battles than most says much about his character. Malozemoff plays to win, whether it is a Granduc project or a golf match. Whatever he does, he does with every bit of energy and intellect he has. He plays fairly, but he plays hard, and he has never been heard to say, "Oh well, it's just a game."

Malozemoff is thorough. Nothing annoys him more than a half-baked opinion based on a superficial reading of the facts. His fierce attack in demolishing such opinions seems sadistic to some observers, but it does discourage repetitions.

How then does he keep track of the dozen or so large and thorny problems thrust at him any and every day? As do most good executives, Malozemoff does it by giving each problem his entire and concentrated attention at the time it is laid before him. His mind does not drift. It digs in, it analyzes, it looks for a conclusion.

Such a conclusion may not always come quickly. Malozemoff has at times maddened his associates by leaving a question dangling for weeks until he has examined it in minutest detail, but once he reaches a conclusion, it is just about set in concrete. When he has fixed on a goal, he drives toward it, and it takes a tremendous effort to cause him to change direction, a quality that has more than once landed him in deep trouble. Nor does he adjust the amount of his interest to match the degree of importance of whatever he is doing. He goes all out in everything. As an industry associate once said, "Never invite Plato into a deal unless you want him in all over, because that's the way he'll be".

Malozemoff drives himself unmercifully, but he is plagued by an uneasy feeling that he is never quite able to arouse his associates to a similar pitch, except when an emergency clearly calls for it. As a result, he feels an urge to be in touch with, and on top of, everything. He does not really want things arranged this way, but that's the way they keep working out.

For a man so devoted to coldly intellectual analysis, Malozemoff is surprisingly sentimental, and his extraordinary loyalty to his old friends has been demonstrated repeatedly. In some people, he has inspired fear; in others, affection; but in nearly everyone who knows him, he has inspired respect. Owing to his sentimental streak, he may want to be liked, but reason tells him that, in his position, it is more important to be respected. What he and his associates in Newmont have done in achieving such growth for the company since 1954 cannot but inspire respect. Even the size and complexity of their mistakes call for it. The real test is now beginning, to see if Malozemoff and the team he is building can maintain the growth of the past and successfully manage the resulting structure. The key members of that team are worth meeting at this point.

Quite typical of today's executive is Jack E. Thompson (no relation to W. B. Thompson), executive vice president, who

came to Newmont in 1960. Jack Thompson was born in Central
City, Nebraska, and began his collegiate studies at Northwestern
University. After spending two years in a pre-law course,
Thompson decided it was not for him. A year of hard work in
construction for a navy contractor bent his interest toward engi-
neering, and he finished his studies in mining geology at the
Colorado School of Mines.

In the spring of 1946, Thompson went to Cuba under contract with Standard Oil of New York, but in 1947 the company
asked him to transfer to the Middle East, whereupon Thompson
resigned. For several years thereafter, he worked on the design
and construction of various chemical complexes for producing
rayon, acid, carbon bisulfide, and other chemicals in Cuba, Venezuela, and Colombia.

All in all, Jack Thompson spent fourteen years in Latin
America, including Cuba and South and Central America. During that time, he built several chemical plants, was general manager of two of them, and acted as a mining and investment consultant for a number of wealthy families in Cuba and Venezuela,
including at one time the mayor of Caracas. Thompson has been
through four revolutions, one in Colombia, one in Venezuela,
and two in Cuba.

During Thompson's last assignment in Cuba, Castro succeeded in his revolution, and Thompson concluded that it
might be better to seek employment elsewhere. At about the
same time that Jack was reaching this conclusion, a friend of his,
Franklin Dewey, who was an old schoolmate of Plato Malozemoff's, mentioned to Plato that he knew of a good man for Newmont.

On the same day that a cable from Malozemoff reached
Thompson in Cuba asking him to come to New York for an in-
terview, Ché Guevara had called on Jack to ask if he would be
interested in heading up the mining department of Castro's Re-
forma Agraria. Though not interested, Thompson told Guevara
he'd have to think it over. The next day Thompson was on his
way to New York, where he accepted a job with Newmont.

After Thompson's return to Cuba, Guevara called on him again to ask about the government job offer. Jack begged off, but recalls that Guevara, a most pleasant and intelligent man, expressed great regret that Thompson could not continue working in Cuba and said that a small matter of differing opinions on politics could cause great disagreement between nations, but should not work such hardship on individuals who could otherwise be friends and partners.

When it came time to leave Cuba, the Thompsons found that they could arrange transportation for themselves but for none of their household belongings. In fact, for some months, there had been no shipments at all out of Cuba to the United States. At his wife's suggestion, Jack went around to see if Guevara might be able to help. He was ushered in by a guard complete with cigar, beard, and machine pistol. Guevara remembered Jack and welcomed him at once. Learning of the difficulty, Guevara called his secretary and issued Jack a pass for himself, his family, and anything he wanted to take out with him. "If anyone ignores that," said Guevara, "he'll go to the wall."

Assigned by Newmont in 1960 to assist in developing new projects, Jack Thompson, at 39 years of age, became assistant to the president in 1964 and was promoted to vice president, development, in 1967. He was elected to the Newmont board of directors in 1969, and, following Roy Bonebrake's retirement at the end of 1971, Thompson was elected Newmont's executive vice president. He is also a director and/or an officer of nearly every Newmont subsidiary or affiliated company, such as Magma, Newmont Exploration Limited, Newmont Oil Company, Idarado, Sherritt Gordon, Foote Mineral, O'okiep, Tsumeb, and Bethlehem Copper. Thompson is also a member of the Advisory Board of the Chemical Bank of New York. His outside interests include organization in 1971 of the Minerals Industry Educational Foundation, an association interested in furthering development of mining and metallurgical engineers through scholarships and grants to qualified students in U.S. mining colleges.

At the head of Newmont's operating staff is David O.

Pearce, vice president for operations, already introduced in the section on O'okiep in Chapter 5. At O'okiep, Pearce became assistant general manager in 1957, general manager in 1959, and a director of the company in 1961. Since 1968, he has been managing director of O'okiep and Tsumeb.

In February, 1968, Pearce moved to New York to become vice president for operations of Newmont, and in May, 1971, he became a Newmont director. Pearce is also an officer and director of several of the Newmont subsidiaries. For example, he is president and director of Idarado, Carlin, Granduc Operating Company, and Newmont Services Limited and is vice president and director of Magma Copper Company, as well as several of the smaller companies. Actually, Dave Pearce is much less proud of these titles than he is of the fact that he is regarded by the members of the operating staffs of the several Newmont subsidiaries as the operator's voice in the New York office. Pearce has experienced the same problems they face, and will work to whatever extent is required to help solve them. He knows the difference between a good job and a bad job when he visits a mine, and he distributes praise or blame accordingly and impartially.

Newmont's exploration programs in the United States, Canada, and Africa are now directed by Richard D. Ellett, who became manager of exploration in June, 1972. A native of Seattle, Washington, Ellett is a graduate in geology of Washington State University. He has worked as a geologist for Patiño in Bolivia, for Newmont and later for National Lead in the southwestern United States, and for Utah Construction Company in Australia. In April, 1970, Ellett became president of Newmont Mining Corporation of Canada Limited, a post that he retains in addition to his present position with Newmont in New York.

In the financial area, Newmont's treasurers, while of differing capabilities, all have been men of exceptional dedication, loyalty, and integrity. Beginning with Henry Dodge, the list includes Gus Mrkvicka, William T. Smith, Walter Schmid, and Gordon Bell. Gordon had assumed the position of treasurer in October, 1969, when Walter Schmid retired, but he had hardly more than settled into the job, when his life was ended in a tragic automobile accident on March 26, 1971. A native of Texas, and educated at Columbia University, Gordon Bell was a fine-looking man of winning personality and with a sincere interest in, and affection for others. An employee of Newmont in the accounting department since 1949, Bell, as the new treasurer, had an excellent career opening before him in the expanding operations of Newmont. On his death, his associates lost a good friend, and the company lost a competent executive.

The post of treasurer is now held by Edward P. Fontaine. An energetic, forceful man of thirty-six, Fontaine came to Newmont from Mobil Oil Corporation, where, in a period of about ten years, he filled various financial positions in the international and North American divisions of that company, most recently as assistant treasurer. A holder of bachelor's and master's degrees in business administration from the University of Michigan, Fontaine has the training, the experience, and the capabilities to fit into, and to further the Newmont approach to flexibility in financing.

The position of controller for Newmont is held by Harry Van Benschoten. Before joining Newmont in December, 1967, at the age of 39, as assistant treasurer, Harry had been associated with Price, Waterhouse & Company, and before that with Pogson, Peloubet & Company, both independent public accounting firms. However, in his activities with these firms, Harry was closely associated with the financial accounting aspects of a number of international mining companies. The move into Newmont's financial department was thus a natural one for him.

Newmont's legal affairs are handled by a group of five lawyers headed by Richard B. Leather, who joined Newmont in 1970 at the age of 38. As chief counsel, Leather replaces Roy Bonebrake, who retired January 1, 1972. Leather came to Newmont from the New York law firm of Chadbourne, Parke, Whiteside and Wolff, where he had substantial responsibilities for several mining company clients. Leather is also secretary of the corporation, and a director of a number of its subsidiaries.

International legal affairs of Newmont are handled by Jacques Leroy, 47 years of age, who has appeared in this history in connection with the lead and zinc mining activities in Morocco and Algeria and the operations of Newmont Overseas Petroleum Company, also in Algeria.

These are some of the men who will be making history for Newmont in the months and years to come. To attempt to forecast precisely what this group will do with the opportunities that will present themselves would involve so many intangibles as to render such speculations meaningless. However, the demonstrated competences and enthusiasms of this essentially youthful staff justify the conclusion that the standard of Newmont performance is hardly likely to be any poorer than it has been. There are, however, a number of tangible elements in the outlook that permit some general conclusions as to the company's course in the next few years.

First, an evaluation of Newmont's future can best be made by considering the present Newmont financial philosophy as it has evolved under Malozemoff's leadership. Subtly, though none the less definitely different from that of his predecessors, this philosophy has developed into Newmont's present approach to financing new ventures.

Obviously, Newmont is now willing and able to consider such financing on a scale that previous administrations could not, or would not, consider. True, McNab carried Magma into the $100 million San Manuel project, but, as has been made clear, Newmont would have no active part of that venture at that time.

Also, Newmont traditionally did not borrow the funds for new ventures. Necessary advances to subsidiaries came out of cash flow or proceeds of sale of part of Newmont's stock portfolio. Only the subsidiaries borrowed such additional funds as were necessary, and Newmont has not had to guarantee such borrowings by its subsidiaries. The only exceptions to this were Newmont's guarantee of a small loan by Western Nickel in 1954 and the guarantee of part of the Palabora loans.

It is one of the more interesting aspects of large-scale financing that a subsidiary of a company with an excellent credit rating, such as Newmont, can borrow a large sum without a guarantee by the parent company. However, this is true only if the lenders feel confident that the parent will stand back of the subsidiary. In other words, the lenders are willing to grant this freedom to a company only if they are sure that the company will not take advantage of it. For example, Newmont did not have to guarantee the loans to Sherritt Gordon because Fred Searls and Franz Schneider assured the Morgan Bank that Newmont would back up Sherritt, and the bank knew that their word was as good as a guarantee.

Another example of loans to a subsidiary without the parent company's guarantee was the financing of the Similkameen project. This could have been done by establishing a fixed debt structure that would have been both burdensome and unnecessary. J. O. Cumberland, who was then the vice president, finance, of Newmont, was instrumental in developing alternatives, one of which was an effort to obtain financing in Japan, where Similkameen's copper concentrates are being smelted. An investigation disclosed that the procedures required were too complicated and the cost too high, in terms of insurance, interest, and required purchases of Japanese equipment. Incidentally, there is no longer anything cheap about Japanese heavy equipment, including the price.

Turning then to the U.S. Export-Import Bank, Cumberland was able to arrange a loan for Similkameen of up to $9 million at 6 per cent interest, of which only about $7 million was taken down, with the requirement that the equipment purchased with the funds must be from the United States. However, the major loan was for $38 million in Canadian funds obtained from the Canadian Imperial Bank of Commerce at a floating interest rate, which was originally 8.5 per cent, but which has run as low as 6.5 per cent. The Canadian loan, payable from May 1, 1973, through 1980, was broken down into a general purpose loan of $28.2 million, an equipment-purchase loan, matching the Ex-Im funds, of $7 million, and a working capital loan of $3.5

million. Both these loans were obtained without a Newmont guarantee. However, Newmont supplied $26 million to Similkameen in equity funds. Cumberland, who was also active in other Newmont financing programs, resigned in 1972 to accept the position of executive vice president of Commonwealth Oil Refining Company.

Elsewhere, Newmont has used preferred stock issues, as in acquiring Magma; leasing arrangements and advances, as in financing Granduc and Atlantic Cement; advances to, or contributions to capital of, subsidiaries; revolving credits from banks; and private placements of loans. Although a public bond issue has been considered in recent years, the financial climate did not make such an issue attractive. However there was arranged in mid-1972 an issuance of tax-free revenue bonds in Arizona amounting to $30 million to help finance Magma's pollution-control program.

Not until the last year or two, however, has Newmont had to do any substantial borrowing on its own. To help finance the Magma expansion, its air-pollution-control program, and other new projects still pending, Newmont itself has had to arrange to borrow about $130 million, and its long-term debt, on a consolidated basis, grew to over $200 million as of the end of 1971. In doing this, Newmont has sought, and achieved, an extraordinary flexibility in financing new ventures by short- and medium-term loans and has thus avoided incurring the conventional long-term-public-debt obligation with its usual rigidities and restrictions in the indenture. This was done in the belief that this ordinary type of debt structure is not advisable for mining companies, whose profits often fluctuate widely with shifting metal prices and the state of supply and demand for metals. Massive repayments of debts can usually be effected during periods of high prices, if repayment terms are not fixed on a long-term basis. Contrariwise, accommodation to the borrower in times of low profitability might occasionally be required.

As a result of its flexible and imaginative approach to financing, Newmont has in the last fifteen years or so been able to finance, or participate in the financing of, a grand total capital investment of $696 million, which was made up by a combination of equity financing by Newmont and/or other participating companies, by borrowings by the subsidiaries and, to a lesser extent, Newmont, and by advances to, or contributions to capital of, the subsidiaries. Newmont's direct contribution to this total capital investment of $696 million was about $165 million. The remainder represents loans or equity supplied by others.

In addition to the $165 million, Newmont invested another $33 million in acquiring holdings in Cassiar Asbestos Corporation, Foote Mineral Company, and Highveld Steel and Vanadium Corporation Limited in South Africa. The investments in Foote and in Highveld have not, thus far, justified the hopes that were at one time held for them.

Newmont sees greater promise in its most recent investment, as of this writing, which was the purchase from Sumitomo Metal Mining Company Limited of a 22 per cent interest owned by that company in the common stock of Bethlehem Copper Corporation Limited, a copper mine in British Columbia. Newmont paid $26.6 million for the 1.4 million shares, approximately their market value at the time of the sale. A part of this sum was paid in cash and the remainder in notes.

Attracted by Bethlehem's ore reserve position rather than by current performance alone, Newmont regarded the Sumitomo offer as a chance to buy a stake in one of the larger copper ore reserves in North America. Bethlehem's J-A orebody in British Columbia contains around 600 million tons assaying about 0.50 per cent copper, and the company has 20 per cent of the adjacent Valley Copper orebody which measures over 900 million tons of like grade. Some form of combined operation of the two properties seems likely.

While awaiting development of this larger potential, Bethlehem is mining smaller orebodies that yield a production of 50 to 55 million pounds of copper annually at a low cost, and a net income running between $5 and $10 million annually over the last five years.

Aside from a greater flexibility in financing demonstrated in recent years, a second element of difference in the present Newmont philosophy lies in Malozemoff's greater use than his predecessors of technical, economic, and financial analysis of a given project. Such analysis has not always guaranteed success, as we have seen, but it has notably supported certain projects, such as Sherritt Gordon's nickel refinery and Magma's copper refinery, that would not have been undertaken had not careful analysis indicated their feasibility and profitability.

Such activity requires availability of a larger staff than Newmont employed previously. Prior to Malozemoff's advent, each Newmont executive had his own area of operations, and there was virtually no staff on which to call for help, except for certain geologists who were skilled in economic evaluations. Newmont's executives can now call on a New York staff of thirteen men competent in the areas of geology, geophysics, mining, metallurgy, mechanical and electrical engineering, mineral economics, computer technology, taxation, transportation, marketing, and of course, law and finance. Even so, most large mining companies would still consider Newmont's staff unusually small for the job it has to do.

A third, and a most significant difference between Newmont's present and its past philosophy is its reaction to the fundamental and sweeping changes that have shaken the world in the last fifty years. Even as late as the 1930's, an international mining company could still regard the world as its oyster, and the only problem was getting the desired part of the oyster ahead of the competition.

Although Spain, Russia, and China were ruled out by political upheaval, the rest of the world was wide open to exploration and development. In evaluating a mineral prospect one had only to consider tonnage, grade, and production economics. Few men ever thought in terms of "political risk." The colonial powers had their foreign possessions well in hand, and one knew where he stood in dealing with them. Despite some unrest in South America, foreign mining companies were still working easily in a well-established pattern of exploration, concession, development, and production, worn smooth by years of practice. Only a few ineffective "agitators" ever used the word "exploitation."

This happy freedom really was lost in the explosion of World War II, although the fact of its loss did not become fully apparent for some years. In Africa, one does not deal with the French, the Belgians, or the British any more. One deals with the governments of newly named countries, unknown previously, such as Zaire, Botswana, Zambia, Tanzania, and Lesotho. In South America, and in much of the rest of the world, one faces a new breed of socialistically and nationalistically inclined military governments or dedicated Communists, but seldom the middle-of-the-road, business-oriented governments with which one dealt in the past.

Without attempting to argue whether this change is good or bad, it is a fact of life today, and it is interesting to observe that regardless of ideologies, or lack of them, the drives motivating all these new controlling elements are essentially the same. These new countries, or the new governments in old countries, want and demand a greater share in the proceeds of their mines and oil wells. These governments and their peoples are aware of the economic opportunities their natural resources represent, and they are determined to hang onto as much as possible of the riches they see in prospect.

The ultimate stage in this drive is, of course, outright seizure, with or without compensation. The major examples have occurred in Mexico, Peru, Chile, Algeria, the Congo, and Libya, and the list is likely to grow even longer. Newmont lost in this way its interest in Algerian oil wells and zinc mines, receiving in return only a fraction of the value of these properties.

The result of all this since World War II, has been a proliferation of laws, taxes, quotas, exchange controls, licenses, mixed-company structures, profit sharing plans, employee participation plans, expropriations, and other schemes, all aimed at reducing the outflow of funds from the natural resource country. In addition, the developing countries want not only mining and smelting of metals at home, they now insist on refining and semifabricating as well, regardless of the dictates of economics. Is this development good or bad? Again, begging the question, it appears here to stay, and must be accepted as the new fact of life.

Consequently, along with the old simple criteria, Newmont today must consider all these new political, social, and economic problems in evaluating a mineral prospect, even in North America. The result is that a particularly complex set of regulations, or an unstable government, or a potentially explosive political situation may automatically rule out a prospect, no matter how attractive it may be otherwise. Recently Newmont has regretfully declined even to look at two inviting prospects in the Middle East, largely because of "political risk." Years ago, Fred Searls would have jumped at them. In fact, he spent a good deal of his time and Newmont's money in a search for prospects in the regions where these two properties are located. Would he still go after them today? He would be tempted, as Newmont is, but the realities of today's world would bind him as firmly as they do his successor.

However, Newmont still selects its investments on the basis of all the available facts (although there are now many more to consider), plus the final essential, though indefinable, element of judgment. Intuition, though still a factor, is certainly of lesser importance than it may have been in former years. As do all mining companies, Newmont looks for properties capable of a high yield despite fluctuating metal prices, and one must agree that at least in some properties Newmont has been more successful than most others in this effort.

Newmont's objective in making a new investment (which quite obviously it has not always achieved) is to secure an eventual after-tax return of at least 20 to 25 per cent. Some of Newmont's investments have paid at a rate vastly higher, but these are the exceptions that a company cannot be sure will be available in the future. Nevertheless, a mining company has to have a generous rate of return just to keep going, let alone grow. The greater the production, the faster the ore reserves will be depleted, and replacing these reserves becomes more difficult and expensive each year.

In contrast, a manufacturer needs only to set up depreciation reserves to pay for replacing his equipment. He can shift his place of business at will and buy his raw materials wherever he can get the best price. A manufacturer can get by nicely at a rate of return of around 8 per cent, and if his rate of return goes much higher, he becomes a phenomenon like IBM, Xerox or Kodak.

Not so in mining. High rates of return are commonplace in mining because they have to be. Dependent on exploration and acquisition for growth, highly capital-intensive in nature, inherently risky in development, mining requires high rates of return to attract new investment capital and generate capital for succeeding ventures as existing ores become depleted. Through aggressive exploration, acquisition of new mining interests, sound technical staff work, and an effective financial policy, Newmont has succeeded remarkably well in maintaining a high growth rate and better-than-average profitability.

In Newmont's short-range picture, two negative elements are the prospective large capital investment for air pollution control in Arizona, and a possible oversupply of copper on world markets. The former holds little hope of profit, but the latter may not develop. Also, there are problems connected with bringing new properties such as Granduc and Similkameen into full production, and there will be continuing large expenditures for exploration in the shrinking number of politically stable areas of the world. Countering these problems will be greatly increased cash flows from Magma and from other investments that have yet to mature fully.

Finally, there is the Newmont portfolio of mining and petroleum stocks which has a market value at this writing of approximately $318 million and which is available to help with new financing. On the average, this portfolio pays a safe return of 4 to 5 per cent, although most of the items in it, considered on a Newmont cost basis, pay at a substantially higher rate. All in all, it would seem that the short-term outlook for Newmont is one of growth, although the earnings may fluctuate owing to changing metal prices.

What of the long-term outlook for Newmont? Unless some subsequent management changes Malozemoff's policy, Newmont will continue to operate in the natural resource field. Such a parameter is wide enough to give Newmont plenty of latitude in diversifying, if it wishes to, away from too great dependence on a single market. Having done quite well in the minerals and petroleum field, there seems no real need, as yet at least, for Newmont to consider expanding into unrelated fields, such as service or manufacturing.

Newmont will no doubt continue its past policy of aggressive exploration in a search for new mineral properties, or for new ventures that it can initiate or in which it can participate. As this record has shown, Newmont has not always tried to run, or to own, everything itself. In fact, some of Newmont's most rewarding ventures have been in combination with one or more partners.

In a way, Newmont's exploration programs begin right in the home office in a search for men qualified to direct such programs in terms not only of geologic skill but of sound judgment as well. To Malozemoff, nothing is more important in evaluating an associate than the quality of good judgment, that indefinable element that cannot be taught, but that makes a man do what turns out to have been the right thing, sometimes without his knowing quite why he has done it. It is a point of pride with Newmont that, more often than not, Newmont's men have exercised better than average judgment in choice of properties whose grade, location, physical features, or other characteristics have enabled Newmont to develop a producing mine that gives the company an edge on its competition. Thompson, Ayer, Searls, and Malozemoff, all have looked for, and have acquired men of ability and sound judgment for Newmont, and the record these men have made speaks for itself.

Thus far in this history there has been presented a picture of Newmont largely as a company that explores for, finances, and develops new mining and petroleum properties. The reason for this emphasis on exploration and development is that it is essentially what Newmont has done throughout its history. True enough, Newmont has managed many of its properties, but it has done so largely through the local managements of subsidiary companies. Whoever undertakes to write the history of Newmont's next fifty years is going to find his task somewhat more complicated.

The reason is quite simple. In the past ten years especially, Newmont has acquired so many new properties, expanded old ones, engaged in so many new investments, that it is encountering somewhat the same sort of problem that has made life so difficult in recent years for the larger conglomerates. In other words, management of the complex Newmont network is not the relatively simple matter it used to be. Obviously, Newmont is no Gulf and Western Industries, and fitting together the interests of O'okiep and Magma is by no means as much of a problem as meshing the interests of, say, New Jersey Zinc and Desilu Productions. Nevertheless, when one has to combine into one optimally profitable whole, the activities of eleven owned or managed companies and fifteen or twenty additional interests or large investments, one definitely has a problem. Newmont is going to find this problem of management increasing, especially if it attempts to maintain past growth rates, which, knowing Malozemoff, one is likely to conclude that it will.

To sum up, then, Newmont, at its half-century mark, enjoys an outlook in which the positive elements substantially outweigh the negative. The latter elements are the problems of air pollution control, adjustment to the environment, development of new methods of financing, development of new management techniques, shifting supply-demand patterns in copper and other products, ever-decreasing availability in the world of more obvious mineral opportunities in areas of political and economic stability. All are formidable problems, but all are susceptible of resolution. The overwhelming positive element lies in the resources of minerals, men, and money on which Newmont can draw in meeting these problems.

In terms of total ore reserves, number of producing properties, gross assets, borrowing power, technical competence, management skills, any measure you like, Newmont, today, is stronger and better equipped than at any time in the past. In progressing from Thompson through Ayer and Searls to Malozemoff, Newmont has lost its initial flamboyance and much of its one-time engaging unpredictability, but it has gained vastly in power, scope, resources, and productivity in its industry throughout the world. Yet, in all this, Newmont has never lost the kind of close-knit, family feeling with which it began.

Objectives have changed, methods have changed, the people have changed, but the company's essential character remains unaltered. Newmont is the continuing creation of a group of talented individuals who work, argue, plan, and build together in what sometimes seems like "confusion worse confounded," but which more often than not achieves a brilliance of outcome that a rigid, highly structured management could not match.

The explantion of how Newmont has retained this small-family quality is really quite simple. From the very beginning, from W. B. Thompson on down the years, the men involved in Newmont have enjoyed themselves. They liked the way the company operated, and each succeeding group has grown up in the same affection.

This history has recorded fifty years of a mixture of successes and failures, false starts and solid achievements, wide disagreements and strong unanimity, but in all this record there has been no occasion to deal with real bitterness or malice. Whatever they did, or did not do, whatever struggles they had, the men and women who made this history obviously had the time of their lives while living it.

If, on perusing this chronicle, you have become aware of

the pleasure these people have felt in struggling against obstacles and overcoming them, you will have understood not only what the company is, but also what this history has been trying to say about Newmont Mining Corporation.

NEWMONT MINING
CORPORATION
Appendices

APPENDIX A

Fifty Years of Financial Growth

The growth of Newmont Mining Corporation in its first fifty years can be traced, in part, by examining the growth in market value and dividend income shown by 100 shares of Newmont common as first offered to the public in 1925. In addition, Newmont's growth in terms of yearly net income, book value, and stock-market value can perhaps best be evaluated by a comparison of Newmont with the performance of four major copper companies over the same period, using 1925 as a base.

Chart No. 1 shows the market value of 100 Newmont shares, first offered to the public in 1925 at a price of $40 per share. If one had bought 100 of the 130,000 shares then offered, and had neither bought nor sold any Newmont shares thereafter, he would now own 4,330 shares of Newmont, obtained through stock dividends and stock splits. For his original $4,000, this investor would have, at the end of 1971, shares that had a market value of about $123,000, yielding $4,503.20 per year in dividends at the current rate.

This chart shows the dramatic rise in market value of Newmont's shares during the early 1920's from the original price of $40 per share and their equally sharp decline after the Crash. Newmont reached its peak at $236 per share in 1929, followed by a drop to its all-time low of $3.875 in 1932 during the Depres-

Chart No. 1: **Market Value of an Original 100 Newmont Shares**
1925-1971 Year-end (in thousands of dollars)

sion. In the chart, only year-end prices are graphed; therefore the highest and lowest values do not appear.

A secondary peak was reached in 1936, reflecting the company's success in developing income from gold mining. The "Roosevelt Depression" of 1937 and 1938 forced the stock market down and, with it, the price of Newmont shares.

A relatively flat period followed during World War II, but a strong upward trend developed from 1948 on, owing to increased income from base-metal mines and from foreign operations. An increase in ownership of Magma Copper Company in 1962 and the fruition (successful development) of several other important investments, including Southern Peru Copper Company, Palabora Mining Company, and others, sparked further advances in the mid-sixties. Naturally enough, the course of the stock market had an important influence on the price trend of Newmont shares, but the company's continued growth had an even greater influence.

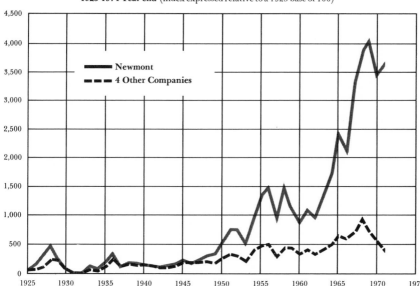

Chart No. 2: **Market Value of Copper Industry Companies**
1925-1971 Year-end (index expressed relative to a 1925 base of 100)

Newmont

4 Other Companies

Chart No. 2 shows the effect of this growth factor on the total market value of Newmont's common shares at year-end in each case. For comparison, the total stock-market values of four other major copper companies are shown. All values are reduced to 1925 as a base.

It is true that only in the last decade or so could Newmont properly be considered one of the major copper companies, for it is only in that period that as much as 85 per cent of its income has been derived from copper. Nevertheless, all through its history, Newmont has been heavily, if indirectly, involved in copper mining both in the United States and abroad; the comparison is therefore believed to be valid.

Except for the 1920's, when the market responded enthusiastically to Newmont's promise and its sharply increased net income, the stock market regarded the companies under study as about equal until 1950. Then Newmont's net income began to increase more sharply than the others (*See Chart 4*) under the influence of O'okiep's and later Tsumeb's earnings.

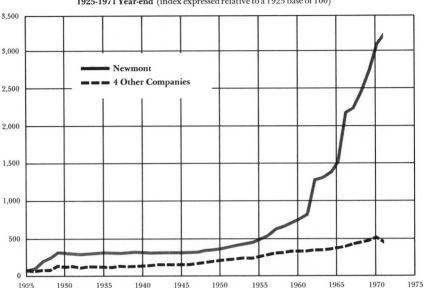

Chart No. 3: **Book Value of Copper Industry Companies**
1925-1971 Year-end (index expressed relative to a 1925 base of 100)

Thereafter, the price of copper stocks fluctuated in accord with fluctuating copper prices, although the continued growth in Newmont's book value and total net income brought about a much steeper rising trend in its market value than was enjoyed by the other copper companies.

Chart No. 3 shows essentially a steady growth in book value (i.e., the net cost of the underlying assets) of the four copper companies used for comparison. Their growth almost exactly paralleled Newmont's rise in book value until 1956, with the exception of the early years up to 1930, when Newmont's book value climbed steeply, owing to reinvestment by Newmont of its stock-trading profits in such growth stocks as Texas Gulf Sulphur, Kennecott, Hudson Bay Mining and Smelting, and others.

After 1956, Newmont's book value increased sharply in comparison with that of the copper companies because of new investments in gas transmission companies and growth in value of assets such as O'okiep and Tsumeb. Sherritt Gordon came

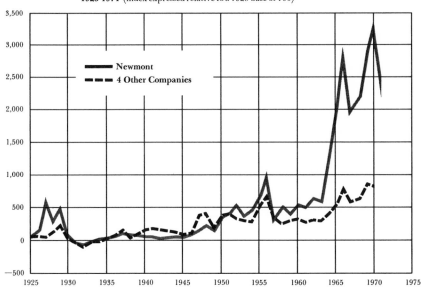

Chart No. 4: **Yearly Earnings of Copper Industry Companies
1925-1971** (index expressed relative to a 1925 base of 100)

— Newmont
■ ■ ■ 4 Other Companies

along in 1952, Southern Peru in 1955, but book value jumped
ahead in 1962, when Newmont acquired 80.6 per cent of
Magma. Investment in several domestic and foreign enterprises
pushed the book value up sharply in 1966. Growth since then
has been steady, with continuing investment at Magma, Similkameen, Granduc, and other locations.

Chart No. 4 shows a comparison of yearly net income between Newmont and the total of the net incomes of the four
other major copper companies studied in preparing Charts 2
and 3. All income figures are reduced to 1925 as a base.

In the 1920's, Newmont's earnings trend far outshone that
of the other four companies largely because of W. B. Thompson's successful stock-market operations. Thereafter, until about
1957, Newmont's net income roughly paralleled that of the
other copper companies, with the major exception coming
in 1956, when, as previously mentioned, sharply increased dividends from O'okiep and Tsumeb, together with high metal

prices, pushed Newmont's net income that year to a new record. Newmont and the other companies then saw their net incomes decline sharply in 1957 due to falling metal prices.

Metal prices rose somewhat in 1958, but wavered uncertainly through the next six years. Newmont's income trended higher than the others, owing to steadily increasing dividend income. Then, in 1964, Newmont's net income took off in a surge caused by dividends received at double the rate of 1962. Groundwork for the 1964 earnings climb was laid in 1962 when Newmont acquired 80.6 per cent of Magma. In February, 1963 Magma paid off the R.F.C. loan with a loan from Prudential that allowed it to resume dividend payments, which added $3,700,000 to Newmont's income in 1964, $6,746,000 in 1965, and $10,288,000 in 1966.

Net income, restated in 1968 for the years from 1952 onward to include Newmont's equity in undistributed earnings of companies of which it owned 50 per cent or more, jumped from $14,006,000 in 1962 to $25,344,000 in 1964. Thereafter, under the impetus of high metal prices in foreign and domestic markets, dividends from O'okiep, Tsumeb, and Magma set new records.

In 1966, Carlin came into full production and paid $3,000,000 in dividends. Southern Peru returned the last of Newmont's advances in 1966 and paid an initial dividend, of which Newmont's share was $4,010,000. Palabora declared its first dividend in 1966, of which Newmont received $280,682. In 1967, Newmont's share of the Palabora dividend came to $2,812,000. All in all, 1966 was one of Newmont's best years, and net income hit a new high of $63,706,000.

The next year, net income plunged to $43,313,000, owing mainly to an eight months' strike that shut down the U.S. copper industry from July 15, 1967, until March 1968. Newmont was harder hit than the other companies.

Rising metal prices strengthened net income into 1969. That year Newmont acquired 100 per cent ownership of Magma, and net income shot up once more. In 1970, under the in-

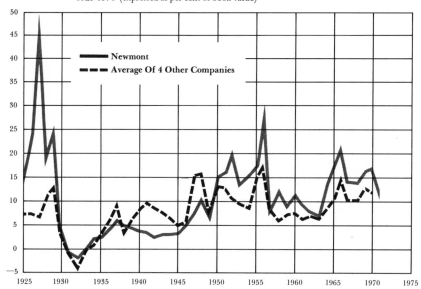

Chart No. 5: **Profit Earned on Book Value by Copper Industry Companies 1925-1971** (expressed as per cent of book value)

fluence of very high foreign and domestic prices, Newmont's net income set still another new record of $72,005,000. Declining metal prices cut net income in 1971 to $54,520,000. The graph for the net income of the other four companies was not extended into 1971 because heavy write-offs taken that year by some of them, reflecting expropriation losses, would completely distort the comparison.

Chart No. 5 shows that the per cent of profit earned by Newmont on its book value has varied widely from the rate of profit of the other copper companies during only three periods. The first such period was in the 1920's, when Newmont's profits from trading skyrocketed, and its only operating expense was for salaries, offices and exploration. These factors made Newmont's per cent of loss somewhat lower than the others during the loss years of the 1930's. It is interesting to note that the other copper companies were able to increase their own per cent of profit during World War II, whereas Newmont was not. The sharp increase in per cent of profit enjoyed by Newmont in the

Chart No. 6: **Dividends per Year on an Original 100 Newmont Shares 1925-1971** (in dollars)

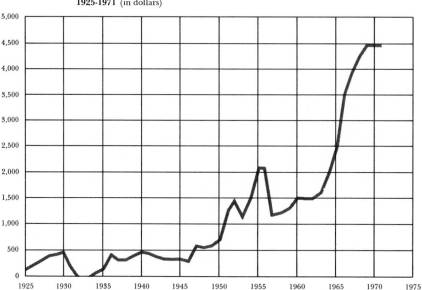

decade of the 1950's was again owing to the substantial contributions of O'okiep and Tsumeb, when high-grade ores and high metal prices in foreign markets enabled unusually high profit margins. It is also interesting to note that, even in 1970, which was a record year for most copper companies because of high prices, the percentage of profit for all the copper companies was lower than in previous years of high metal prices.

Chart No. 6 shows the growth in dividends paid on 100 Newmont shares, bought in 1925, and reflecting, of course, stock dividends and splits through the years. No dividends at all were paid in 1932 and 1933. Income from Newmont's gold-mining operations in the United States and Canada enabled an increase in dividends in the mid-1930's to nearly the $4.00-per-share level reached in 1929 and 1930. Declining profitability of gold mining, owing to increasing costs in the years just prior to World War II, and the flat earnings during the war brought about a reduction in dividends for that period. Thereafter, owing to increased investments in base-metal mining and to a generally rising trend in metal prices that resulted in substantially increased income, first from the foreign properties and later from Magma, dividend payments were rapidly increased. A temporary reduction became necessary in the late 1950's and early 1960's, but the upward trend resumed after the acquisition of 80.6 per cent of Magma in 1962.

Clearly the record of growth established by Newmont has been derived from alert and continuing investment in new mining properties, some of which (O'okiep, Tsumeb, Southern Peru, Carlin, and Palabora) have been extraordinarily rewarding. The largest investment, Magma Copper Company, is currently in the process of developing its full potential, which should begin to be expressed in 1973 and 1974. Magma's contribution, together with that from other properties still in the consideration stage, can be expected to maintain Newmont's growth, although it would be perhaps too optimistic to expect a continuing growth as dramatic as that portrayed in these charts.

APPENDIX B

Officers, Board Members, and Management

DIRECTORS AND OFFICERS

Board of Directors

Plato Malozemoff
  President and Chairman of the Board
Roy C. Bonebrake
  Retired, formerly Executive Vice President and General Counsel, Newmont Mining Corporation
Gordon H. Chambers
  Private Investor, Philadelphia, Pennsylvania
Frank Coolbaugh
  Consultant, Denver, Colorado
Lewis W. Douglas
  Honorary Chairman of the Board, Southern Arizona Bank & Trust Co., Tucson, Arizona
Wesley P. Goss
  Chairman, Magma Copper Company, San Manuel, Arizona
Christian Hohenlohe
  Attorney, Washington, D.C.
André Meyer
  Senior Partner, Lazard Frères & Co., New York
William B. Moses, Jr.
  Chairman, Massachusetts Financial Services, Inc., Boston, Massachusetts
David O. Pearce
  Vice President
Walter P. Schmid
  Retired, formerly Treasurer, Newmont Mining Corporation
Stuart F. Silloway

Director, Investors Diversified Services, Inc., Minneapolis, Minnesota
Jack E. Thompson
  Executive Vice President

Officers

Plato Malozemoff
  President and Chairman of the Board
Jack E. Thompson
  Executive Vice President
Richard D. Ellett
  Vice President, Exploration
David O. Pearce
  Vice President, Operations
Robert J. Searls
  Vice President, Australia
Oscar F. Tangel
  Vice President, Research and Development
Richard B. Leather
  Secretary and General Counsel
Thomas G. Watkinson
  Assistant Secretary
Edward P. Fontaine
  Treasurer
Winthrop T. Parker III
  Assistant Treasurer
Harry Van Benschoten
  Controller and Assistant Treasurer
Walter E. Baker
  Assistant Controller
Robert F. Boyce
  Assistant Controller, Taxes

Charles F. Tiller
Assistant Controller, Metallurgical Accounting

MANAGEMENT PERSONNEL

Newmont Mining Corporation

STAFF:

L. A. Cassara
Marketing
D. J. Christie
Project Engineering
P. J. Crescenzo
Assistant Chief Engineer
J. R. Denny
Systems Engineer
M. E. Emerson
Minerals Economist
J. P. Fitz-Gibbon
Counsel
C. G. Freeman
Mining Engineer
J. C. Keenan
Senior Mining Engineer
D. M. Koogler
Assistant to Vice President of Operations
J. L. Leroy
International Counsel
R. H. Ramsey
Assistant to the President
E. H. Tucker
Chief Engineer
H. W. Volkman
Purchasing and Insurance

CONSULTANTS:

M. D. Banghart
Mining Engineer, 9 Monmouth Avenue, Claremont, Cape Province, Republic of South Africa
R. C. Bonebrake
Counsel, 300 Park Avenue, New York, N. Y. 10022
F. W. McQuiston, Jr.
Metallurgical Engineer, 230 Kaanapali Drive, Napa, California 94558
W. K. Pincock
Mining Engineer, 4858 East Broadway, Tucson, Arizona 85711

Newmont Exploration Limited

Research Laboratory, 44 Briar Ridge Road, Danbury, Connecticut 06811
Dr. A. A. Brant
Director, Geophysical Department
R. D. Macdonald
Director, Metallurgical Department

Exploration
B. S. Hardie
Box 368, Elko, Nevada 89801
H. J. Steele
Box M. San Manuel, Arizona 85631

Newmont Mining Corporation of Canada Limited

R. D. Ellett
President, 1610–25 King Street West, Toronto 105, Ontario
R. F. Sheldon
Vice President, Exploration, 1230-355 Burrard Street, Vancouver 1, British Columbia
J. Drybrough
Director, 906-211 Portage Avenue, Winnipeg 2, Manitoba

Newmont Oil Company

Dr. R. S. Moehlman
President, 1135 Capital National Bank Building, Houston, Texas 77002
J. L. George, Jr.
Vice President, Exploration
E. D. Whitis
Vice President, Production

Newmont Proprietary Limited

R. J. Searls
Managing Director, AMP Tower, 535 Bourke Street, Melbourne, Australia 3000

Dr. J. Claveau
Chief Geologist, 14 Chancery Hill
Road, Singapore

Newmont Services Limited

J. M. Petty
General Manager, Dawn, Idarado
and Carlin, Idarado Mining
Company, Ouray, Colorado 81427
E. M. Craig
Resident Manager, Dawn Mining
Company, Box 25, Ford, Wash-
ington 99013
P. N. Loncar
Resident Manager, Idarado Min-
ing Company, Ouray, Colorado
81427
J. D. McBeth
Resident Manager, Carlin Gold
Mining Company, Box 672, Elko,
Nevada 89801
N. Gritzuk
Vice President and Manager,
Granduc Operating Company,
890 West Pender Street, Van-
couver 1, British Columbia
R. S. Mattson
Resident Manager, Granduc
Operating Company, Stewart,
British Columbia
J. H. Parliament
Executive Vice President, Simil-
kameen Mining Company Lim-
ited, 890 West Pender Street,
Vancouver 1, British Columbia
A. F. Bissett
Resident Manager, Similkameen
Mining Company Limited,
Princeton, British Columbia

Newmont South Africa Limited

V. Vellet
Manager of Exploration, 99
Eloff Street, Johannesburg, South
Africa

M. Crichton
Counsel, A. Livingstone & Co.,
P.O. Box 3920, Johannesburg,
South Africa

Magma Copper Company

W. P. Goss
Chairman
W. H. Burt
President and Chief Executive
Officer, San Manuel, Arizona
85631
J. S. Wise
Vice President and General Man-
ager, San Manuel Division, San
Manuel
D. J. Buckwalter
Assistant General Manager, San
Manuel
E. K. Staley
Manager, Superior Division, Su-
perior, Arizona 85273

Maluti Holdings Limited

V. M. Reinecke
General Manager, 117 Commis-
sioner Street, Johannesburg, South
Africa

**O'okiep Copper Company Limited
Tsumeb Corporation Limited**

D. O. Pearce
Managing Director, New York
J. L. Leroy
Secretary, New York
G. R. Parker
General Manager, O'okiep, Na-
babeep, Cape Province, Repub-
lic of South Africa
J. P. Ratledge
General Manager, Tsumeb, South
West Africa

Acknowledgments

NEWMONT HISTORY

Since 1921, Newmont Mining Corporation has been involved in approximately ninety-five major projects or investments. In addition, the company has investigated, or actually explored, several hundred prospects, and has considered and rejected thousands more. To evaluate all these, and to describe accurately those worthy of inclusion in this book, required extensive study of the Newmont files and innumerable conversations with present and former Newmont staff members and with friends in other mining companies as well.

Of particular help with information on the Thompson era and the 1920's were: Earle K. Currie, Warren Publicover, Franz Schneider, and William T. Smith.

More recent data on Newmont and its subsidiaries came also from these men and from the following present Newmont group employees: David J. Christie, Eileen Clohosey, John C. Keenan, Jacques L. Leroy, P. Malozemoff, David O. Pearce, Robert J. Searls, Oscar F. Tangel, Jack E. Thompson, Eugene H. Tucker, Ruth Vanderpoel, Henry W. Volkman, John S. Wise, and Aasine Womack. Robert S. Moehlman supplied much of the material on Newmont Oil Company, and James P. Ratledge, Peter Philip, and G. Söhnge provided technical and historical data on Tsumeb.

The following retired Newmont or affiliate employees were most generous in their help: M. D. Banghart, Roy C. Bonebrake, John Drybrough, Wesley P. Goss, Frank W. McQuiston, Jr., Fred A. Scheck, and Walter P. Schmid. Former employees, now with other companies, who were most helpful were: Julian O. Cumberland, Robert B. Fulton, John S. Livermore, and Francis E. Rinehart.

Carroll Searls was a mine of information on Newmont, Empire Star, and his brother, Fred, but, unhappily, Carroll died before seeing the completed manuscript.

Friends in the industry outside Newmont who contributed greatly were: Eldon L. Brown, Francis Cameron, James Douglas, John K. Gustafson, Forrest G. Hamrick, and Robert P. Koenig.

Finally, acknowledgment is due the patience and thoroughness of Amy Oddo and the other girls in the File Department, and of Filomena De Ciantis and Virginia McCleerey, who typed the several drafts and sorted out the seemingly endless revisions and corrections.

Bibliography

NEWMONT HISTORY

Hermann Hagedorn, *The Magnate.* New York: The John Day Company, Inc., 1935.

A. B. Parsons, *The Porphyry Coppers.* New York: American Institute of Mining Engineers, 1933.

Arnold Hoffman, *Free Gold: The Story of Canadian Mining.* New York: Holt, Rinehart & Winston, 1947.

Isaac F. Marcosson, *Metal Magic.* New York: Farrar, Straus & Cudahy, 1949.

Harvey O'Connor, *The Guggenheims: The Making of an American Dynasty.* New York: Covici, Friede, 1937.

Theodore Gregory, *Ernest Oppenheimer and the Economic Development of Southern Africa.* Cape Town, South Africa: Oxford University Press, 1962.

James M. Morley, *Gold Cities.* Berkeley, California: Howell-North Books, 1965.

William Haynes, *The Stone That Burns.* New York: Litton Educational Publishing, Inc., Van Nostrand-Reinhold Company, 1942.

INDEX